Christoph Brosinsky

On power system automation: A Digital Twin-centric framework for the next generation of energy management systems

Ilmenauer Beiträge zur elektrischen Energiesystem-, Geräte- und Anlagentechnik (IBEGA)

Herausgegeben von
Univ.-Prof. Dr.-Ing. Dirk Westermann
(Fachgebiet Elektrische Energieversorgung) und
Univ.-Prof. Dr.-Ing. Frank Berger
(Fachgebiet Elektrische Geräte und Anlagen)
an der Technischen Universität Ilmenau.

Band 33

Christoph Brosinsky

On power system automation: A Digital Twin-centric framework for the next generation of energy management systems

Universitätsverlag Ilmenau
2023

Impressum

Bibliografische Information der Deutschen Nationalbibliothek
Die Deutsche Nationalbibliothek verzeichnet diese Publikation in der Deutschen Nationalbibliografie; detaillierte bibliografische Angaben sind im Internet über http://dnb.d-nb.de abrufbar.

Diese Arbeit hat der Fakultät für Elektrotechnik und Informationstechnik der Technischen Universität Ilmenau als Dissertation vorgelegen.

Tag der Einreichung:	10. Januar 2022
1. Gutachter:	Univ.-Prof. Dr.-Ing. Dirk Westermann (Technische Universität Ilmenau)
2. Gutachter:	Hon.-Prof. Dr.-Ing. Rainer Krebs (Siemens AG)
3. Gutachter:	Prof. Dr. Peter Palensky (Delft University of Technology)
Tag der Verteidigung:	8. Juli 2022

Technische Universität Ilmenau/Universitätsbibliothek
Universitätsverlag Ilmenau
Postfach 10 05 65
98684 Ilmenau
https://www.tu-ilmenau.de/universitaetsverlag

ISSN 2194-2838
ISBN 978-3-86360-266-6 (Druckausgabe)
DOI 10.22032/dbt.54812
URN urn:nbn:de:gbv:ilm1-2022000425

„Essentially, all models are wrong, but some are useful."

George Edward Pelham Box[1]

[1] George E. P. Box and Norman R. Draper, *Empirical Model-building and Response Surface*. NY, USA, 1986.

Danksagung

Die vorliegende Arbeit entstand während meiner Tätigkeit als wissenschaftlicher Mitarbeiter am Institut für Elektrische Energie- und Steuerungstechnik der Technischen Universität Ilmenau. An dieser Stelle möchte ich mich bei allen Menschen bedanken, welche mich bei der Erstellung dieser Arbeit begleitet haben.

Zunächst möchte ich mich bei meinem Doktorvater Herrn Univ.-Prof. Dr.-Ing. Dirk Westermann für die persönliche Betreuung und Förderung innerhalb der vergangenen Jahre bedanken. Das hohe Maß an Freiheit und entgegengebrachtem Vertrauen hat die Zeit am Fachgebiet Elektrische Energieversorgung stark geprägt. Zudem möchte ich Herrn Prof. Dr. Peter Palensky und Herrn Hon.-Prof. Dr.-Ing. Rainer Krebs für die Übernahme der Gutachtertätigkeit, sowie für das ermunternde Feedback zur vorliegenden Arbeit bedanken.

Auf meine fachliche und persönliche Entwicklung hatten vor und während meiner Tätigkeit als wissenschaftlicher Mitarbeiter an der Technischen Universität Ilmenau viele Personen Einfluss. Bei Dr.-Ing. Steffen Prinz und Dr.-Ing. Michael Malsch möchte ich mich für das Entfachen der Begeisterung für die vielfältigen Themen der elektrischen Energietechnik bedanken. Dr.-Ing. Anne-Katrin Marten danke ich für die Förderung und Unterstützung zu Beginn meiner Tätigkeit als wissenschaftlicher Mitarbeiter und die Gelegenheit im Projekt Kriegers Flak Combined Grid Solution mitzuwirken. Bei Dr. Philipp Sauerteig bedanke ich mich für die konstruktiven Gelegenheiten einen Einblick in die Arbeitsweise von Mathematikern zu gewinnen.

Ich bedanke mich ebenfalls für den intensiven Austausch bei meinen Kollegen Dr.-Ing. Steffen Schlegel, Dr.-Ing. Florian Sass, Dr.-Ing. Robert Schwerdfeger, Dr.-Ing. Teng Jiang, M.Sc. Tom Sennewald, M.Sc. Franz Linke, sowie bei allen anderen Kollegen und Kolleginnen, mit denen ich zusammenarbeiten durfte.

Für die Ermöglichung des prägenden Aufenthalts als Gastwissenschaftler am Campus Risø der Danmarks Tekniske Universitet (DTU) bedanke ich mich bei meinen Betreuern Alexander Maria Prostejovsky (Ph.D.), Kai Heussen (Ph.D.) und Mattia Marinelli (Ph.D.).

Weiterhin gilt mein Dank allen Projektpartnern aus Industrie und Wissenschaft für die vielzähligen Anregungen und Inspirationen in diversen Diskussionen.

Nicht zuletzt möchte ich mich bei meiner Familie, meiner Partnerin und meinen Freunden für die Unterstützung, Motivation und Geduld in den vergangenen Jahren bedanken. Ohne euch wäre die Fertigstellung dieser Arbeit nicht möglich gewesen.

Kurzfassung

Das elektrische Energiesystem befindet sich in einem umfangreichen Transformationsprozess. Durch die voranschreitende Energiewende und den zunehmenden Einsatz erneuerbarer Energieträger werden in den nächsten Jahren in Deutschland viele konventionelle Kraftwerke stillgelegt. Diese Veränderungen führen unter anderem zu einer verringerten Systemträgheit und somit zu einer zunehmenden Anzahl komplexer dynamischer Phänomene im elektrischen Energiesystem. Der zukünftig häufigere Betrieb des Energiesystems nahe an den physikalischen Systemgrenzen führt des Weiteren zu einer erhöhten Störanfälligkeit, zu erhöhter Betriebsmittelauslastung und zu verkürzten Zeitspannen, um auf unvorhergesehene Störungen zu reagieren. Infolgedessen wird die Aufgabe das elektrische Energiesystem zu betreiben und zu führen immer anspruchsvoller. Insbesondere wenn Reaktionszeiten erforderlich werden, welche die menschlichen Fähigkeiten übersteigen, sind die zuvor genannten Veränderungen intrinsische Treiber hin zu einem höheren Automatisierungsgrad in der Netzbetriebs- und Systemführung.

In der vorliegenden Arbeit wird ein Framework für eine neuartige modulare Leitsystemarchitektur vorgestellt, welche als zentrale Komponente einen sogenannten Digitalen Zwilling enthält. Im Konzept des Digitalen Zwillings werden Informations- und Betriebstechnologien vereint, so dass dieses als Grundlage für zukünftige Leitsystemarchitekturen und Anwendungen der Leittechnik herangezogen werden kann. In Ergänzung zu den bereits vorhandenen Funktionen moderner Netzführungssysteme unterstützt das Konzept die Abbildung der Netzdynamik auf Basis eines dynamischen Netzmodells. Um eine realitätsgetreue Abbildung der Netzdynamik zu ermöglichen, werden zeitsynchrone Raumzeigermessungen für die Modellvalidierung und Modellparameterschätzung herangezogen. Dies erhöht die Aussagekraft von Systemsicherheitsanalysen, sowie das Vertrauen in die Modelle mit denen operative Entscheidungen generiert werden.

Durch die Bereitstellung eines validierten, konsistenten und wartbaren Datenmodells auf der Grundlage von physikalischen Gesetzmäßigkeiten und während des Betriebs gewonnener Prozessdaten, adressiert der vorgestellte Architekturentwurf mehrere Schlüsselanforderungen an moderne Netzleitsysteme. So ermöglicht das Framework einen höheren Automatisierungsgrad des Stromnetzbetriebs und den Einsatz von Entscheidungsunterstützungsfunktionen bis hin zu vertrauenswürdigen Assistenzsystemen auf Basis kognitiver Systeme. Diese Funktionen können die Betriebssicherheit erhöhen und stellen einen wichtigen Beitrag zur Umsetzung der digitalen Transformation im Energiesystembetrieb, sowie zur erfolgreichen Umsetzung der Energiewende dar. Das vorgestellte Konzept wird auf der Grundlage numerischer Simulationen untersucht, wobei die grundsätzliche Machbarkeit anhand von Fallstudien nachgewiesen wird.

Abstract

The electrical power system is in the process of an extensive transformation. Driven by the energy transition towards renewable energy resources, many conventional power plants in Germany have already been decommissioned or will be decommissioned within the next decade. Among other things, these changes lead to an increased utilisation of power transmission equipment, and an increasing number of complex dynamic phenomena. The resulting system operation closer to physical boundaries leads to an increased susceptibility to disturbances, and to a reduced time span to react to critical contingencies and perturbations. In consequence, the task to operate the power system will become increasingly demanding. As some reactions to disturbances may be required within timeframes that exceed human capabilities, these developments are intrinsic drivers to enable a higher degree of automation in power system operation.

This thesis proposes a framework to create a modular Digital Twin-centric energy management system. It enables the provision of validated and trustworthy models built from knowledge about the power system derived from physical laws, and process data. As the interaction of information and operational technologies is combined in the concept of the Digital Twin, it can serve as a framework for future energy management systems including novel applications for power system monitoring and control, which consider power system dynamics. To provide a validated high-fidelity dynamic power system model, time-synchronised phasor measurements of high-resolution are applied for validation and parameter estimation. This increases the trust into the underlying power system model as well as the confidence into operational decisions derived from advanced analytic applications such as online dynamic security assessment.

By providing an appropriate, consistent, and maintainable data model, the framework addresses several key requirements of modern energy management system architectures, while enabling the implementation of advanced automation routines and control approaches. Future energy management systems can provide an increased observability based on the proposed architecture, whereby the situational awareness of human operators in the control room can be improved. In further development stages, cognitive systems can be applied that are able to learn from the data provided, e.g., machine learning based analytical functions. Thus, the framework enables a higher degree of power system automation, as well as the deployment of assistance and decision support functions for power system operation pointing towards a higher degree of automation in power system operation. The framework represents a contribution to the digital transformation of power system operation and facilitates a successful energy transition. The feasibility of the concept is examined by case studies in form of numerical simulations to provide a proof of concept.

Table of Content

Keywords: Digital Twin, Energy Management System, Dynamic Security Assessment, Dynamic State Estimation, Moving Horizon Estimation, Power System Operation, Assistance System

Introductory note: Parts of the research results presented coherently within this thesis have been published before in international publications. For the sake of transparency and scientific probity these publications are listed in the appendix A.6.

1 Introduction and Motivation

A digital transformative phase currently embraces all kinds of businesses and provides new opportunities for deployment of new technologies. The evolution of technologies, such as cloud computing or artificial intelligence, influences enterprises and organisations all around the world [1]. Several transformative innovations and novel technologies also impact the energy sector, where emerging technologies strongly influence the energy transition (ger: *Energiewende*). In conjunction with the deregulation of electricity generation and unbundling of the market from transmission and distribution tasks, the complexity of power system operation continues to increase. This implies the need for higher flexibility in operation of the transmission (and distribution) networks, whether through novel technological approaches or the optimisation of market driven processes.

A rising number of converter-interfaced renewable generation units is integrated into the power system. The volatile nature of renewable energy sources results in a rising effort for transmission and distribution system operators to maintain system security and causes high redispatch costs. Additionally, many conventional power plants are already decommissioned or will be put out of service within the next decade. This leads to a reduced power system inertia and results in a lacking voltage and frequency support [2]. The decreasing system inertia in turn, leads to an increasing number of dynamic phenomena and higher gradients in the rate of change of frequency (ROCOF) [3]. In addition, fast controllable converter coupled generation units, voltage source converter high voltage direct current (VSC-HVDC) transmission links and flexible AC transmission system (FACTS) devices add further complex dynamic phenomena to the electrical power system. In consequence, the time span in which effects of disturbances can be effectively mitigated decreases, while the susceptibility of the electrical power system to the impacts of severe faults increases [2], [4].

The electrical power system can be split into transmission system, sub-transmission and distribution system [5]. An illustration of a modern interconnected power system is given in Fig. 1.1 (inspired by [6]). The transmission system operated by a transmission system operator (TSO) at the highest voltage levels (typically, 220 kV and above), interconnects all major generating stations and main load centres in the electric power system. The sub-transmission system transmits power in smaller quantities from the transmission substations to the distribution substations, from where large industrial customers are commonly supplied. A clear demarcation between sub-transmission and transmission system is not always given [5]. Especially in Europe, the transmission and sub-transmission layer are not explicitly considered as separated and are typically located within the same area of responsibility of TSOs. The distribution system, operated by a distribution system operator (DSO) represents the transfer layer, where power to the individual customers is distributed and accounted. As shown in Fig. 1.1 all parts of the system are interconnected electrically, whereas different voltage levels are interfaced by transformers. In addition to the electric interconnection, the observability of the entire power system becomes crucial for the secure operation. To obtain real-time observability of the operational state, a vast number of sensors, data processing infrastructure, as well as modern information and communication technologies (ICT) are implemented (see red connections in Fig. 1.1). Together, these constitute a so called cyber-physical system.

The enhancement of power system monitoring and control systems is addressed by researchers worldwide. Functional requirements of future control centre applications have been documented by EPRI in [7]. The authors of the report identify the trend towards faster than real-time simulations and look-ahead strategies to forecast system states, in close to real-time operation. In [8], Wu et al. reviewed the past, present, and likely future of the control centre functions and architectures and identify a gap between today's applied and available technologies. Zhang et al. presented a vision of the next-generation monitoring, analysis, and control functions. They identify a bottleneck towards future control centres in the sequential computing, which is still applied in most of today's control centres, instead of capable parallel computing infrastructure [9]. Future Supervisory Control and Data Acquisition systems are envisioned in [10, 11]. The authors state, that these will evolve into dynamically adaptive systems, with modular software architectures, considering vendor independent standards. Tomsovic et al. propose a framework for real-time control, but emphasise the high reliability demands of such a system, which enforces a conservative design [12]. In summary, these researchers make use of the leverage the developments in information technology (IT) and operational technology (OT), which the so called trend towards IT/OT convergence promises [13], [14].

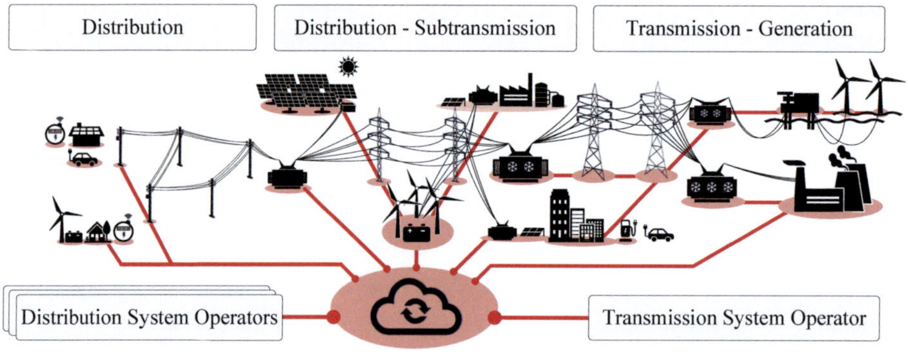

Fig. 1.1: Illustration of the power system from a cyber-physical perspective (inspired by [6])

New approaches to visualize the power system state in the control room have been thoroughly discussed by Hauser [15] and Schneiders [16]. While Hauser provides an integral approach to display the global power system state, Schneiders proposes a hierarchical multilayer display of the global system state, to illustrate deviations from intended conditions. Both approaches base on the idea of conveying essential information through an adaptive visualization, considering the human cognitive barrier [17].

A key characteristic of future power system control systems is the requirement to observe, monitor and analyse power system dynamics. Dynamic phenomena cannot be properly observed by conventional network control technology with the classic SCADA approach. Therefore, new monitoring and control applications, such as PMU based WAMS are deployed. With the deployment of PMU based wide area monitoring systems (WAMS) in control rooms, appropriate streaming data visualisation became a new field of research. PMUs enable a system wide time synchronised observation of the system frequency as well as complex voltage and current quantities. Thus, system states and dynamics can be obtained almost in real time. Hence, modern energy management systems (EMS) applications often relate to synchrophasor data streams. Giri et al. describe a framework of control room analytics and visualization for enhanced situational awareness [18]. Brahma et al. [19] apply phasor measurement units (PMU) data streams for real-time identification of dynamic events. Crouser et al. [20], identify a problem related to the pace at which the incoming streaming data are changing: Streaming data visualizations require to meet the capability the human intellect and its perceptual limits, while adapting to a limited data processing speed.

1.1 Trends in Power System Automation

Due to the increasing complexity of system operation, the automation of the power system is often involved in ongoing discussions. Generally, the term automation describes a system's ability to operate, act or self-regulate itself independently, i.e. without human intervention [21]. The automation of power system processes and control enables real-time operation of sensitive tasks, which must be completed far below the human reaction time (e.g., frequency and voltage control, protection and redispatch). A higher level of automation enables efficient processing of repetitive tasks, which saves time and resources. Automation furthermore helps to reduce the number of human errors, improves productivity, raises efficiency, and reduces costs at the same time [22]. To maintain quality of supply, while reaching economic objectives taking several constraints into account, leads to an intrinsically motivated need for automation [23]. The characteristics and challenges imposed by the operation of future power systems are summarised in Tab. 1.1.

Tab. 1.1: Summarised characteristics of future power systems (see also [24, 25])

Aspect	Characteristics
Inertia	Decommissioning of rotating machinery (i.e., conventional power plants) and a high share of converter-coupled generation units lead to high frequency gradients, due to lower system inertia [4, 26].
Uncertainty	Significant temporal and spatial uncertainty, due to weather dependent power in-feed by renewable resources, a higher number of extreme weather events [27], and unbundling of energy marketing and transmission tasks in Europe impose higher uncertainty.
Observability	Ubiquitous sensors, WAMS and smart meters lead to a rising observability [28, 29].
Controllability	ICT connected controllable assets, enable new system control approaches. Power electronics interfaced flexible loads and generation (e.g., demand response, DER flexibility) raise remote control possibilities [28].
Flexibility	Flexible power system equipment to control power flows, such as VSC-HVDC-links or FACTS will gain relevance in future. Responsive power electronics interfaced demand and storage management systems add flexibility.
Resilience	Preventive and curative control actions in addition to structural reinforcement and grid expansion increase the resilience of the power system to severe operational conditions or events [30, 31].
Trust	Higher confidence in models by online validation from measurement data.
Coordination	Increased need for horizontal and vertical TSO/DSO coordination.
Communication technology	ICT in the power system allows a high degree of observability and controllability striving towards IT/OT convergence [13].
Automation	Slow evolution towards highly automated or even autonomous systems [22]. System operation tasks, underlying processes and work routines in the control room will change.
System operation in the control room	Control centre software architecture evolves from monolithic to modular [24]. Data analytics and automated processes in the control room will turn data into knowledge. EMS/ DMS functionalities utilise available degrees of freedom, advanced operator decision support and assistant systems implemented.
Asset utilisation	A higher asset utilisation and equipment loading, leads to an operation closer to operational limits.

Automation systems have been an integral part of power system operation and management for several decades. The first step towards power system automation was already undertaken, when load frequency control was incorporated into EMSs [25, 32]. Since then, automated feedback controllers were included also in other parts of the system, especially were reaction times beyond human capabilities are needed [25]. EMS and operational procedures in the control room, however, have largely remained unchanged so far. From the perspective of a rapid technological progress and changing demands for power system control, the question raised why electric power grids and the energy sector did not move towards full automation, like many other industries [25].

Various degrees of automation exist, while manual control by humans contradicts automation. A few possible intermediate levels between manual control and full automation, i.e. autonomous systems are given in [21, 22, 25]. Following the definition of automation and the corresponding classification in the literature, a highly automated power system [25]:

- handles normal operating conditions without manual intervention by humans,
- offers decision support to human operators in alert and critical conditions,
- reduces human workload considerably.

The need for increased real-time information exchange and automation to assist in power system operation has been emphasised in consequence to power system incidents in the past. The primary causes have been unpredicted equipment outages, lacking observability by adequate monitoring systems, and the ability to take control actions [33].

The increasing complexity of power system operation, can cause struggling of human operators with their supervision and control tasks [34]. Furthermore, humans tend to lose their problem solving skills, when subjected to time pressure [35]. Thus, automation systems need to take over more responsibilities and approaches for automated decision support in power system operation are required [25]. While an increasing level of automation has many benefits, it also may cause a detachment of human operators from the system, leaving them in a low awareness state, sometimes referred to as the "*out-of-the-loop*" syndrome [36]. This entails the potential threat of reduced situation awareness and an increased risk of human errors [35, 37]. In conclusion, frequent tasks involving a lot of data processing and analysis are recommended to automate. But the decision to automate automatable decision processes which include human experience and intuition should be split by human reaction time and the capability of humans to react appropriately. For the latter, including decisions relevant to system integrity, partial automation is recommended where necessary. Thus, the final decision-making will remain a task for the human operator. In addition, the operator's role will include defining and adjusting the ruleset for the automation system [25].

Although the necessity of human involvement in system control in presence of automation is questioned at times, a closer look on the extent and aspects of automation reveals the indispensable nature of human intuition when dealing with abnormal situations [25].

An emerging EMS application for the analysis of power system dynamics during operation is online dynamic security assessment (DSA). DSA systems apply dynamic simulations and have been applied in the past usually during the planning and system design phase. Today, it is mandatory for TSOs within the ENTSO-E to assess dynamic stability at least once a year to identify stability limits [38]. In the meantime, the need for analyses based on dynamic system models has moved closer to real-time system operation. Therefore, online

DSA functions are gaining interest among system operators worldwide. This has been also shown by the incident on January 8th 2021, where the continental European synchronous area was separated. The investigating committee concluded to enforce stability margin assessment for operational planning and power systems operation [39].

An approach to combine automatic learning techniques and DSA has been proposed in [40] and discussed before in [41]. The *"automatic operator"*, as Dy-Liacco called the approach, is supposed to run within the control room EMS environment to let the power system itself provide the necessary variations to create a valuable training set. He stated, that the value of conducting DSA studies online *"is that for highly-meshed networks the brink of instability is reached only after an accumulation of several contingencies"* [40]. Thus, the approach takes advantage of the fact that numerically unstable scenarios are very unlikely under normal system conditions. According to the approach, the affiliated automatic learning process runs continuously in real-time, obtaining samples of the operating variables. Those are then further processed by a DSA software in combination with additional routines for selection and execution of simulations for credible contingencies.

Computer aided control of power systems in real-time, has been applied since the 1970s [42], but has been limited by processing speed. Today, affordable high performance computing (HPC), including multicore parallelisation, and specialised computing platforms such as GPUs and FPGAs, pave the way to automated dynamic operator assistant systems [43], [44]. Such a system could generate and execute processes e.g., adapting setpoint settings of controllable devices or switching processes automatically. Besides application of specialised HPC platforms, the research for faster modelling techniques, especially in the time-domain has gained interest in research [45, 46].

Advanced statistical methodologies and learning algorithms, are expected to play an important role in preparation of fully automated process sequences [22, 25]. With suitable algorithms, the conventional operator interaction with the control systems can be tracked, analysed, and optimized. The control system may propose recommendations to the operator to take measures based on best practices in similar situations from the past or based on experiences of other human operators. In consequence, the human operator can be guided through system operation by an assistance system, comparable to an aircraft pilot who has activated the autopilot for certain tasks under his responsibility.

Summarising the research on power system automation, some general trends pointing towards the next generation of EMS can be identified. While the functions of the process, field and substation control level in the high and extra-high voltage networks have already been automated to a large extent [47], developments in EMS are not in line with the developments in the power system. A significant automation potential at the system level in the control centre EMS is still unused [47, 48]. In general, implementations of process automation and new proactive, action-oriented features in the control room are on the rise to support operation of the grid closer to its limits and minimising preventive margins [49]. Considering a rising number PMUs installed in the power system, a *"real-time automated closed loop control from a centralized EMS system becomes viable"* [50].

1.2 Towards a Future Proof Modular Power System Operation System

Operational technology of critical infrastructure can be extended only with high effort and at a great expense. Furthermore, systems present in the industry are often historically grown monolithic architectures. Although reasonable from a historical point of view, these monolithic system architectures are not feasible anymore to provide enough flexibility to adapt to the fast-changing requirements of power system operation. Most technologies and applications running in today's power system control centres were developed decades ago, and only incremental changes in functions have been made from time to time [7]. Due to the high effort and expense, system operators conduct major updates to their EMS environment only once in a decade or when requirements change significantly. Thus, an upgraded EMS needs to address possible future scenarios and utility needs. It needs to be adaptable to customer requirements and applicable to operate the increasingly complex and steadily evolving electric power system.

The main disadvantage of a monolithic system is the direct dependency between the system components, i.e., a failing component can lead to incorrect behaviour or failure of the entire system. Another major disadvantage is their difficult maintenance, which is also related to the interdependence of the system components. This makes it arduous to implement new functions or to react to changes in the requirements imposed by external entities. Furthermore, monolithic architectures imply limited competition among vendors for the sake of an optimal solution of a certain functionality. These drawbacks are of course well understood and newer developments point towards modular adaptable architectures with standardised interfaces. This can be identified as a general trend in the development of architectures for utility software solutions and EMS [23], [51].

Emerging real-time applications in IT, and OT enable new innovative concepts to design and operate cyber-physical systems (CPS), which relate to physical systems integrating computation and communication technology [52]. Within the field of system operation, including EMS/ DMS environments this trend is also referred to as IT/OT convergence, which is embodied in the concept of the *Digital Twin* (DT) [14, 23]. The DT concept is a promising approach, which has been discovered recently as key technology by several industries [53–56]. It is often used synonymous to the term CPS where engineering disciplines are involved or in focus [57]. In general DTs refer to virtual representations of systems, processes or objects, that are able reflect the physical conditions by state abstraction and linking the physical and digital world by sensor data streams [58]. The concept is promising for applications where enhanced observability is required, and future operating conditions and system states need to be predicted. For this reason, the concept of CPS and DTs are also involved in recent discussions about roadmaps towards autonomous systems and concepts to operate power systems automatically [22, 59]. Due to the multitude of possible applications, a uniform definition of a DT is impossible, and what exactly constitutes a DT is difficult to grasp. According to the main inherent characteristics described in the literature, the DT incorporates the following characteristics:

- It comprises an adequate analytic model based on *first principles*, i.e., a digital representation of the real system [56], [60],
- It supports automatic and bidirectional data exchange in real-time [61], [62],
- It enables accurate state reflection of the connected physical object, process, or system [63], [64].

Taking these characteristics into account, a *Digital Twin* can be described as a continuously adaptive digital model, which allows to apply analytic functions to gather new information about the system it mirrors [58]. Thus, the main difference of a DT and a conventional simulation model is the ability to stay synchronous to the connected system. Abstracting the DT concept further, it can serve as the core instance of an "operating system" for any physical system (SystemOS). The SystemOS, as illustrated in Fig. 1.2 collects and processes data, thus it provides a fully comprehensive system state by mapping of real-time data streams to a virtual model. Its layered architecture shares common properties with the open system interconnection (OSI) model [65] applied for IT network communication, the Smart Grids Architecture Model (SGAM) [66], which facilitates interoperability requirements to operate ICT enabled power systems (so called "*smart grids*"), and with the service oriented architecture of the Reference Architectural Model Industry 4.0 (RAMI 4.0) [67].

The generic SystemOS illustrated in Fig. 1.2 sets the motivation for the proposed DT-centric architecture of an energy management system. It provides a future proof interface for applications to operate power systems in a secure and reliable way. It furthermore allows to provide relevant parts as open-source software, rising interoperability and comparability of calculation functions and their results.

By representing a meta-architecture for converging IT/OT applications, the conceptual SystemOS illustrated in Fig. 1.2 provides a framework for modular EMS extensions and customised applications, supporting standardised interfaces for data exchange. Its functions and services provide a software environment for system operation and control, as well as interfaces to all other interacting components, such as sensors and actuators data sources, and application interfaces by standardised communication. These interfaces enable flexible creation and deployment of customised applications. Thus, the architecture of the SystemOS allows to integrate applications independently of a provider in a modular way. A DT-centric EMS build upon the SystemOS architecture, as illustrated in Fig. 1.2 provides several advantages in comparison to available solutions:

- A modular, maintainable, expandable, and customisable software architecture,
- Standardised data models and interfaces for data and model exchange,
- Support of legacy and novel applications by standardised interfaces,
- Possible retrofitting of enhanced assistant functions and decision support applications.

The evolutionary steps of the power system towards a highly integrated cyber-physical system utilising the DT concept as an embedded core function within the EMS are shown in Tab. 1.2.

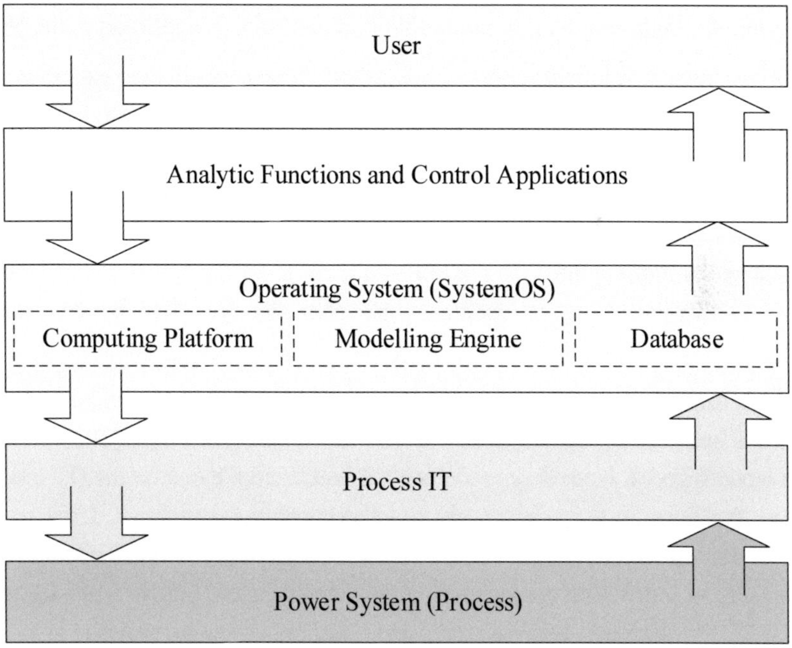

Fig. 1.2: Illustration of the conceptual SystemOS (adapted from [68])

Tab. 1.2: Evolution of EMS from the perspective of cyber-physical systems

Evolvement	EMS Features [69]	Functionality
1st Generation	Hard wired, fully analogue communication	Relevant measurement values are visualised by basic SCADA systems
2nd Generation	IP/TCP based communication	SCADA and state estimation allow quasi-steady-state observability
3rd Generation	Support of dynamic observability	Wide area monitoring and dynamic security assessment tool available
4th Generation	Digital Twin centric architecture	Real-time simulation of high-fidelity models is a core functionality

1.3 Scope and Contributions of the Thesis

In a typical control room, operators rely on many different software applications. The task of sifting through a myriad of SCADA data, weather reports, alarm lists, contingency analysis tables and calculation or state estimator results imposes high demands on the staff [70]. Thus, they must maintain situational awareness under high volumes of rapidly changing data. Novel data sources and tools, such as WAMS and DSA intensify this development.

WAMS integrated into modern control room EMS allow to receive constant feedback of certain power system variables indicating the state of the power system [23]. Furthermore, DSA systems are mandatory in future systems operation according to article 38 of the System Operation Guideline [38]. To validate the required time-domain models online, remains an obstacle for fast implementation. Although, the need for an EMS functionality, which integrates data captured by WAMS to qualify power system models for accurate dynamic security and stability assessment studies has been identified, but does not exist [71].

Within this thesis the potential for automation of energy management and power system operation is investigated. Starting with a review on the state of the art of EMS and power system automation, an innovative modular EMS architecture based on the DT concept is proposed. Preliminaries to implement the novel architecture are analysed. These comprise structured data models (e.g., the Common Grid Model Exchange Standard (CGMES)), and modern SCADA and WAMS communication standards (i.e., IEC 60870-5-104, IEC 61850, IEEE C37.118), which connect data model and measured values are analysed in a requirement analysis.

The trend of EMS development towards the next generation DT-centric EMS is illustrated in Fig. 1.3. Starting from the third generation of EMS (see also Tab. 1.2), which supports dynamic observability of the power system, the next evolutionary step in EMS development towards the DT-centric EMS architecture is anticipated. The proposed DT-centric EMS allows to assess and recommended control actions by subroutines. Therefore, a modelling engine, a so called dynamic digital mirror has been developed, which supports dynamic power system simulations. The proposed modular DT-centric EMS architecture described in this thesis, comprises a high-fidelity simulation model, derived by a model validation routine, which compares observed measurements and simulation model output. The parameters of the model are updated in case a mismatch between observed system state and simulation model output by a suitable parameter estimation methodology. Thereby it enables increasingly precise statements about the system state and the security margin in real-time [69]. It is intended that it can be used for novel grid automation concepts and as a prerequisite for highly automated grid control.

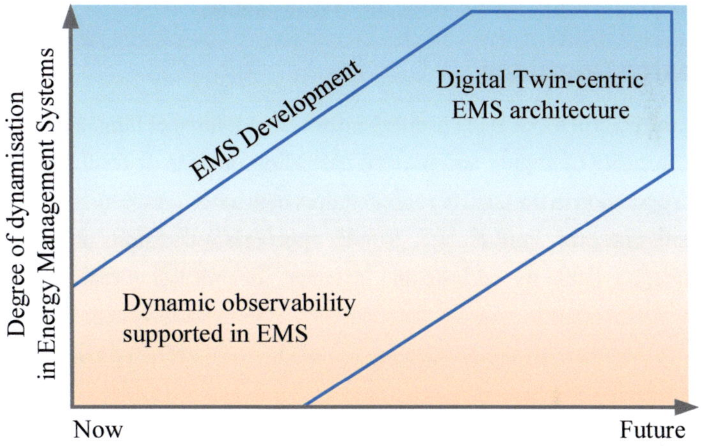

Fig. 1.3: Dawn of the proposed next generation DT-centric EMS

The time-domain simulation engine in connection with the validated DT simulation model, enables accurate state reflection of the observed power system. Thus, the additional benefit of the conceptual DT-centric EMS is, that it enables the accurate estimation of the power system state, and provides a validated model basis, which allows to carry out dedicated power system studies online e.g., a dynamic security analysis. The evaluation of the proposed EMS architecture is conducted by case studies involving numerical simulations in the time-domain. The provided case studies aim towards accurate and reliable DSA.

The contributions of this thesis can be summarised as follows:

- Elaboration of the Digital Twin concept, discussion of its requirements, and its advantages for automation, monitoring and control of power systems,
- Proposal of a novel Digital Twin-centric architecture for the next generation of EMS to enable a higher degree of power system automation,

- Development of a modelling engine which can run steady-state and dynamic power system simulations of hybrid HVAC-HVDC grids,
- Development of a Digital Twin simulation engine, the so called dynamic digital mirror (DDM),
- Development of corresponding interfaces to apply the DDM for online system studies,
- Proposal of a metric to measure the model accuracy,
- Implementation of a suitable methodology for online parameter estimation of time-domain power system models to improve the accuracy of the power system state reflection,
- Demonstration of the feasibility of the proposed architecture by validation of selected central components in numerical case studies.

1.4 Research Questions

The EMS in the control room is the primary information source of human system operators. To maintain security of supply and prevent cascading outages or resulting blackouts, today's system operation in the control room requires new tools to capture real-time dynamics of the transmission grid. Furthermore, to take appropriate decisions and apply remedial actions in time, a reliable model base and instances for processing real-time data streams are required. As the time to react to state changes in the power system is expected to decrease, the development towards a partly automated and coordinated control of the power system is inevitable.

Recently, advanced computation hardware capabilities, specialised simulation platforms and algorithms are available to enable power system computation, analysis, and real-time feedback. A promising approach, which has recently gained much attention is the DT concept. It is suitable to close the prominent gap between existing control centre technology and future needs, rising from new operational requirements. The tools provided can help operators to act faster and apply substantiated decisions based on validated, i.e., trusted simulation models. The thesis envisions the next generation of energy management systems, including dynamic security and wide-area situation awareness, which in turn enable a higher degree of automation up to closed loop system operation. The concept furthermore facilitates the optimisation of power system operation, efficiency and reliability.

Within this thesis the prevailing technologies are examined, and the requirements to enable the DT approach in future EMS are defined. In addition, the feasibility of the proposed concept as a key component of intelligent automation systems to operate electric power systems will be evaluated. To do so, the following research questions have been formulated, which will be discussed in detail throughout this thesis. The answers given within the thesis are summarised in subsection 6.2.

RQ 1. Which functionalities are required to create a future-proof EMS and how can they be achieved?

RQ 2. What is a Digital Twin, and which characteristics distinguish it from conventional models?

RQ 3. What are the key requirements to create a power system Digital Twin for real-time applications?

RQ 4. How could the architecture of a Digital Twin-centric control centre EMS look like?

RQ 5. Which algorithm is suitable for improving the accuracy of the power system model to develop a dynamic Digital Twin?

RQ 6. Can a Digital Twin-centric EMS increase the degree of automation in system operation?

RQ 7. What impact does the new concept have on operational processes?

1.5 Thesis Outline

The thesis is organised in several sections. The introduction and the motivation to the research towards a future proof modular EMS are given in section 1. Here, the scope and the contributions of the thesis are outlined by research questions and objectives to be investigated. Section 2 provides a detailed state-of-the-art analysis, which is concluded by a gap analysis. The methodologies and techniques chosen to design the novel DT-centric EMS framework are described in section 3. The Digital Twin concept and its use cases are investigated, and a definition is given to support the purpose of the thesis. Furthermore, a functional method based on a moving horizon approach is described in section 3. It serves the purpose of dynamic model parameter estimation. In section 4 the requirements and a design approach for a future-proof modular EMS architecture are proposed. Selected components of the proposed DT-centric EMS are validated by case studies in section 5. Section 6 concludes the work, recapitulates the research questions, and summarises the respective answers given throughout the work. Finally, future research directions are pointed out. The bibliography and appendix follow the concluding sections.

2 State of the Art in Technology and Research

All functions required in an EMS are linked to the day-to-day tasks and activities of power system operators. Starting with the presentation of related research projects in Europe in subsection 2.1, subsection 2.2 describes the process of power system operation in accordance with ENTSO-E standards. This includes a review and explanation of basic concepts such as system security and resilience. It also focuses on current processes in the control centre and the tasks of a system operator. Human factors involved in grid control and the task of energy management in control centres, especially the strongly bound concept of situation awareness are presented. Subsection 2.3 presents the current state of available tools and methods for monitoring and control of the power system. The concept of wide area monitoring systems, which are lately deployed to capture network dynamics are explained in subsection 2.4. Subsection 2.5 briefly summarises the state of research regarding assistance systems for human power system operators in the control room. Subsection 2.7 concludes the state-of-the-art analysis resulting in a gap analysis.

2.1 Related Research Activities

This subsection serves to situate the content of this thesis into the context of current research projects. During the preparation of this thesis, the author was intensely involved in several research projects that are closely related to the presented results (especially *DynaGrid Control Center*[2],*HyLITE*[3], and *InnoSys2030*[4]). A central research objective of the projects was, to find solutions for a comprehensive of future power system operation [23, 72, 73]. Special attention was dedicated to the treatment of dynamic phenomena in the power grid and related stability and safety aspects. In comparison to the conventional power system operation based on steady-state models, the next generation of EMS comprises a high-fidelity time-domain model representation. However, among other things, the parameters that determine the dynamic response of the models applied are only known to a limited extent and need to be estimated.

The research projects on European scale which cover related topics are briefly described as follows. The project *ELECTRA* evaluated the need for decision support systems within future control rooms and proposed new scenario-based analytics and visualisations to assist operators. Human factors involved in grid control and energy management in control centres have been considered and evaluated [25].

---

[2] Funded by the German Federal Ministry for Economic Affairs and Climate Action (BMWK grant 03ET7541D)
3 Funded by the German Federal Ministry for Economic Affairs and Climate Action (BMWK grant 0350034C)
4 Funded by the German Federal Ministry for Economic Affairs and Climate Action (BMWK grant 0350036M)

The *GARPUR* project aimed to develop, assess, and evaluate new reliability criteria to maintain power system performance, considering socio-economic aspects and regulatory aspects at a pan-European level.

The aim of the *MIGRATE* project was to seek solutions, which can address the technological and regulatory challenges the power system will face by the massive integration of power electronic devices, today and in future. New real-time monitoring and forecasting solutions have been investigated, to determine the true stability limit of a system as well as control and operation strategies, and protection schemes at transmission system level.

The *PEGASE* project aimed at improving computational tools and algorithms used by TSOs to operate the power system at national level considering the coordination for integrated operation as a part of the European transmission network. In focus were algorithms for state estimation, steady state optimisation, and time-domain simulation. A particular aspect has been to improve power system models, in such way, that observability is maintained in real time for large power systems.

Within the *UMBRELLA* project, a toolbox for TSOs has been developed, to addresses the challenges of power system operation. It considered power systems subjected to high penetration of intermittent and volatile renewables and the associated forecast errors on different time scales.

The project *iTESLA* dealt with the increased need for coordination of operating procedures among transmission network operators. New tools were developed to quantify the distance of a system state from the next security boundary. These considered dynamic security margins derived by risk-based security assessment, and operational degrees of freedom incorporated by flexible transmission equipment, such as HVDC, PST and FACTS.

The Electric Power System Research Institute (EPRI) conducts long term research on control centre functionalities. These consider situational awareness during real-time operation, monitoring, control, and data analysis including synchrophasor applications with focus on human system interface design for advanced grid monitoring and control functionalities. The "*Control Center of the Future*" workshop organised by EPRI in April 2020 is representative for the current interest of system operators in improving the control centre technology. The initiative named the "*Control Room of the Future*" is currently driven forward by a research and development cooperation between the TSO TenneT and EPRI [74].

Research on power system Digital Twins is conducted currently by several research institutions and universities e.g., in the project *Deep Digit* at TU Eindhoven [75] or at the TU Delft [76]. Furthermore, a Task Force of the German VDE ETG investigated the DT concept and its manifold areas of application within the energy sector [77].

2.2 Power System Operation

The operation of the electrical power system is regulated by legal frameworks to maintain security of supply while increasing market competitiveness. The regulations enacted by the European Commission at European level are implemented in national regulations. These comprise the guideline on electricity transmission system operation [38], the Continental Europe Operation Handbook (CE-OH) [78], the ENTSO-E Network Code on Operational Security (NC-OS) [79], and the Network Code on HVDC Connections (HVDC-NC) [80]. Relevant German laws are in addition the EnWG [81], which regulates the legal requirements for electrical grid operation, and the Renewable Energy Sources Act (EEG) [82], which establishes the priority utilisation of renewable energy resources.

2.2.1 Operational Planning and System Operation

The operation of the electrical power system can be regarded as "*a series of control actions which are taken to maintain continuity of service at standard frequency and voltage*" [83]. The time frames of power system operation can be divided into operational planning and system operation. The latter can be further subdivided into close-to-real-time (C2RT), and actual real-time operation. Here, C2RT denotes the planning timeframe below one hour from actual real-time. The operational planning, which comprises all processes to prepare real-time operation, deals with long-, medium- and short-term scheduling, as well as the planning for the future real-time operation [25]. While, long-term and day-ahead operational planning procedures are still mainly carried out in the back office, a growing number of intra-day actions are moving into the control room [34]. However, relevant data from operational planning (e.g., planned outages due to maintenance) are considered in the intraday processes supported by the EMS and real-time system operation. In the Day-Ahead Congestion Forecast (DACF), the demand and generation forecast are used to predict power flows for the upcoming day. This information helps to detect critical situations in the transmission network and initiate appropriate measures to mitigate congestions at an early stage. Usually, this is realised by employing congestion management schemes, confirming unit commitment and acquisition of ancillary services. The predicted power flows often deviate from the ones taking place in real-time operation. Influencing factors that can lead to power flow deviations are, among others, power plant outages, load losses, forecasting errors, or unplanned switching as well as equipment failures. Furthermore, the deregulation of the energy markets obliges the grid operator to purchase ancillary services from the electricity markets, often close to real-time operation. Thus, additional information processing capabilities between grid control centres and electricity market systems are required [34]. Both, the real-time system operation, and the operational planning are illustrated in Fig. 2.1.

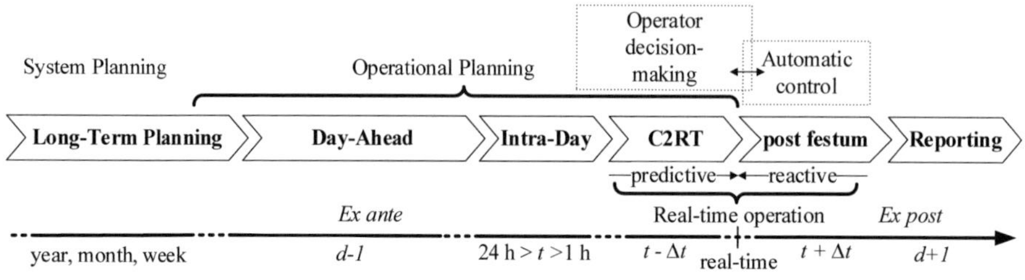

Fig. 2.1: Timeframes of processes related to operational planning and system operation (partly adapted from [84], [85])

2.2.2 Power System Resilience and Operational Security

A general definition of resilience of power systems has been proposed by the IEEE Task Force on Definition and Quantification of Resilience. Following this definition, resilience is *"the ability to withstand and reduce the magnitude and/or duration of disruptive events, which includes the capability to anticipate, absorb, adapt to, and/or rapidly recover from such an event"* [86]. The Cigré WG C4.47 defined power system resilience as *"the ability to limit the extent, severity, and duration of system degradation following an extreme event resilience"* [...] and *"is achieved through a set of key actionable measures to be taken before, during, and after extreme events"* [87]. Thus, the capability of the power system to anticipate, absorb, adapt, or recover from disturbances, comprises operational planning, and the handling of events in real-time operation. To illustrate the interaction of the different mechanisms, and to evaluate the severity and the impact of an event to the resilience, Panteli et al. developed the resilience trapezoid [88]. To legitimate emergency control actions, power system state conditions have been defined. The classic definitions of normal and abnormal power system operating states according to [79, 89, 90] are described in Tab. 2.1. Remaining system operation in the exceptional "*alert*" state is only tolerable in case of availability of appropriate special protection schemes (SPS), which prevent cascading failures when critical contingencies or severe disturbances occur. While the "*normal*" and "*alert*" states are common, the "*emergency*" state is usually reached only after severe disturbances or cascading outages.

It is envisioned, that future intelligent power system operation enables operational states beyond the conventional *N*-1 security criterion (i.e., the power system must remain secure even if an additional device fails) by taking available curative measures into account [91]. SPS are designed to respond upon the detection of single events (e.g., loss of a line) or a particular combination of events, when the power system is not necessarily in an emergency state or close to instability [89]. Addressing the same principle, the North American Reliability Corporation (NERC), coined the term Remedial Action Scheme (RAS) to point out the auxiliary character [92, 93].

Tab. 2.1: The definition of operational states [79, 89, 90] and actions to maintain or to recover normal operation (previously published in [25])

State of operation	Definition of condition	Control actions
Normal	All boundary conditions fulfilled.	No intervention required
Alert	Security-level below limit, boundary conditions still fulfilled	Preventive measures to restore adequate reserve margins needed (e.g., security dispatch, increase of operational reserves, interconnector rescheduling)
Emergency	Boundary conditions violated, countermeasures of alert state insufficient or too slow	Immediate intervention needed (e.g., fault clearing, fast redispatch, load shifting)
In extremis	Boundary conditions violated and system balance disturbed	Drastic measures to contain disruption of the entire system (e. g., load shedding or islanding)
Restorative	System equipment available for restoration	Re-establish sustainable system state (i.e., restart and synchronization of generation units, load restoration, synchronization of areas)
Blackout	(Parts of the) system down	Post-mortem analysis, preparing black start-capable units

Both SPS and RAS allow to operate the power system closer to its physical and technical boundary limits and represent an economic feasible alternative to rising redundancy by construction of additional reinforcing infrastructure [94]. In addition to SPS and RAS, System Integrity Protection Schemes (SIPS) comprise a set of such coordinated and automatic measures designed to initiate the final attempt to stabilize the power system after a large disturbance [89].

As illustrated in Fig. 2.2 for future system operation, curative measures are expected to extend the range of safe or alert system operation. Hence, a former alert or emergency state becomes a *"curative (N-1)-secure"* state, which takes into account fast adjustments in the grid [95]. As a result, the state with previously violated boundary conditions becomes an area of operation with temporarily permissible higher utilisation due to curative measures. An *"alert"* state then only arises when the curative measures are exhausted. These approaches include all degrees of freedom in power system operation, such as load shedding and generation unit curtailment or disconnection, automatic setpoint adaption of HVDC and FACTS devices, switching of shunts and topological changes [96]. As these measures cause complex interactions within the power system and may also endanger system security if not properly designed, the DT-centric EMS allows to evaluate the power system resilience to likely disturbances and to assess such remedial actions in the sandbox online before activation.

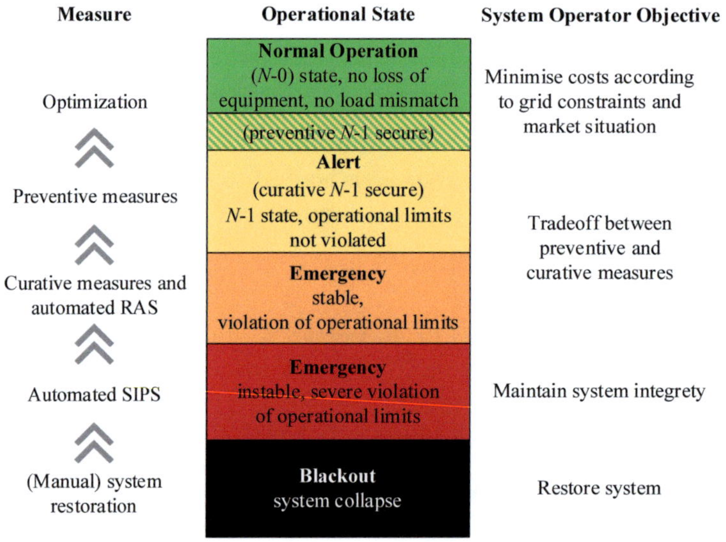

Fig. 2.2: Traditional classification of system operational states according to [97], enhanced by preventive and curative *N*-1 security allowing temporarily increased equipment utilisation closer to physical operating limits

2.2.3 Tasks of the Control Room Operator

From putting infrastructure and components into operation, maintaining their functionality, to bringing the system back on-line after blackouts, humans are involved to varying degrees at all stages of power system operation. Human operators work in an environment, which requires to switch between situations comprising high workload, where quick and accurate reaction is required, and periods of slow pace comprising daily routine checks of the system status, review of operating procedures, and the assessment of activities from the previous shifts [98]. Their duty is to intervene manually when required, considering cost-effective and feasible solutions by first applying no-cost measures before taking more expensive solutions. Human reaction time in power system control is usually considered to be 5 to 15 minutes [99]. The increasing complexity of the power system leads to a rising number of events to handle, which limits this time to react. Therefore, a lot of attention has been put on the development of automated power system control systems, which entail a shift in the role and tasks of human operators in the control room [25]. The daily tasks and the cognitive challenges of human operators in highly automated power systems of human operators in the control room have been analysed in advance, in order to consider those in the design of the novel EMS architecture applying the DT concept.

As illustrated in Fig. 2.3, and listed in Tab. 2.2, power system operators have to cover a range of knowledge areas to maintain safe and efficient grid operation under changing conditions: Resource and demand balancing, transmission system equipment, emergency preparedness, emergency response, contingency analysis and reliability, as well as communication technology and data handling [100]. The need for communication to different internal processes, company subsections or other third parties lead to an increasing number of ICT supported processes, which require strong knowledge on the performed operational tasks. Characteristics of future system operation are a higher uncertainty and consequently less predictable system states, resulting in new approaches for risk management. Hence, the decision itself and the required action must be implemented in less time. Therefore, new automation systems are required which support advanced alarm handling and appropriate reactions on events within a reduced timeframe. Hence, an extensive control room operator training is required in future, to undertake complex analyses in the timeframe C2RT, as well as a training for system state awareness during real-time operation and the handling of upcoming automation systems [101]. With the increasing use of automation systems for the operation of power grids, this new field of knowledge will become more important for grid operators, although it is not yet considered in practice and therefore missing in Fig. 2.3.

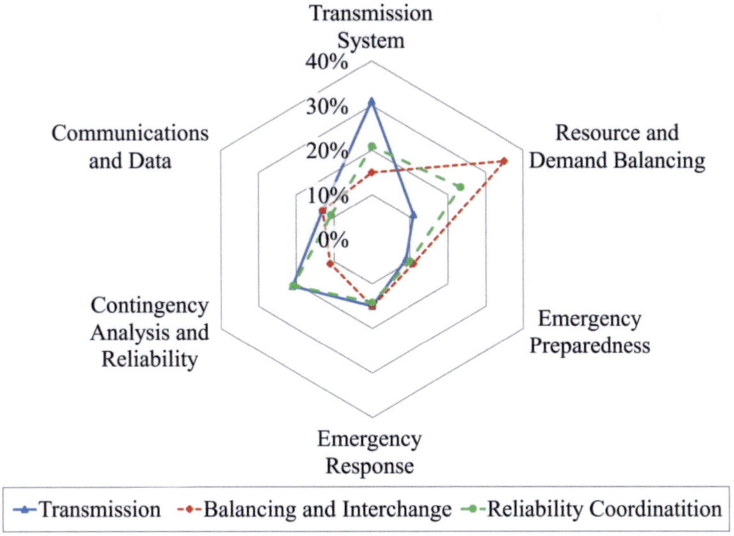

Fig. 2.3: Kiviat diagram representing work domains of operators based on NERC training and certification program [100] (previously published in [25])

Tab. 2.2: Tasks from the NERC operator training and certification process [25], [100]

Category	Tasks
Planning	Define boundaries such as system operating limits or total transfer capabilities
	Define emergency procedures and system restoration plans
	Coordinate and revise generation and transmission maintenance plans
	Identify, communicate, and direct actions against threats and violations
	Implement emergency procedures and restoration plans
	Perform and interpret contingency analysis and dynamic security assessments
	Coordinate field crews and schedule asset maintenance
Monitoring	Calculate and monitor the area control error
	Monitor interconnection reliability and limits to prevent instability and cascading
	Monitor and adjust reactive resources to maintain system voltage within limits
	Monitor and deploy emergency procedures, protective relaying systems and RAS
	Monitor and update telemetry and telecontrol parameters within the control area
Response	Adjust power flow and approve arranged interchange in the system
	Curtail confirmed power interchanges that adversely impact system reliability
	Deploy ancillary services for balancing generation and load considering reliability
	Direct emergency procedures (e.g., load shedding and system restoration)
	Operate the control area to maintain the generation-load-interchange balance
	Conduct switching and develop automated switching regimes
Report-ing	Pursue operational planning and reliability evaluation
	Provide energy accounting and administer inadvertent energy paybacks
	Review generation commitments, dispatch, and load forecasts

2.2.4 Coordinated System Operation and Future Role of TSO and DSO

The presumption of decades, that electric power systems, including grids and plants are natural monopolies, changed towards a trend of deregulation and marked liberalisation. A process which is also known as ownership unbundling [102]. The goal of unbundling ownership by means of separating control of the transmission and distribution grid from the responsibility of generation, comes at the cost of lost vertical economies, but stimulates competition, triggers investments, and accelerates the evolution towards an integrated European energy market [103]. While TSOs are responsible for balance and integrity of the transmission grid, DSOs are primarily concerned with power supply of the contracted consumers, metering, monitoring of power quality parameters, and the provision of non-discriminatory grid access. Furthermore, DSOs are responsible for the congestion management at their respective voltage levels [6, 104]. Since the distributed generation capacity is mainly installed in the distribution system (110-kV voltage levels and below), the responsibility to balance generation and demand of electricity, and for congestion management (Redispatch 2.0) is partly shifted towards DSOs. The changing role of DSOs into active system coordinators [105], requires new data management and exchange schemes between TSOs and DSOs in order to enable active collaboration, utilization of flexibilities and to avoid control conflicts in vertical and horizontal system operation [25]. Coordinated operational planning and system operation emphasising interaction of DSO and TSO requires additional information exchange. Especially the necessary provision and coordination of ancillary services, such as frequency control, system balancing, voltage control, and system restoration, rely on data exchange for coordinated decision processes [106]. An exchange of simulation models and associated parameters between TSOs becomes inevitable in the future to maintain system security [38].

A modern EMS therefore needs to support the Common Grid Model Exchange Specification (CGMES) specified in IEC 61970-600-1 [107] and IEC 61970-600-2 [108]. Furthermore, the exchange of forecasted generation and load data among TSOs and DSOs as specified in the Generation and Load Data Provision Methodology (GLDPM) is required. It contains information required for day-ahead and intraday planning procedures. Following the coordination requirement, a modern EMS supports the related processing and preparation of data automatically.

The need for a mutual real-time data exchange between system operators within the interconnected ENTSO-E system has been proven by the implementation of the European Awareness System [109].

By providing an IT-platform for monitoring and sharing operational information an efficient coordination is implied. Hence, all involved system operators must apply the same data structure [110]. As illustrated in Fig. 2.4 (adapted from [110]), all data sets must be available prior to the start of a new coordination cycle. The proposed DT-based EMS architecture can support to automate this data preparation and exchange procedure.

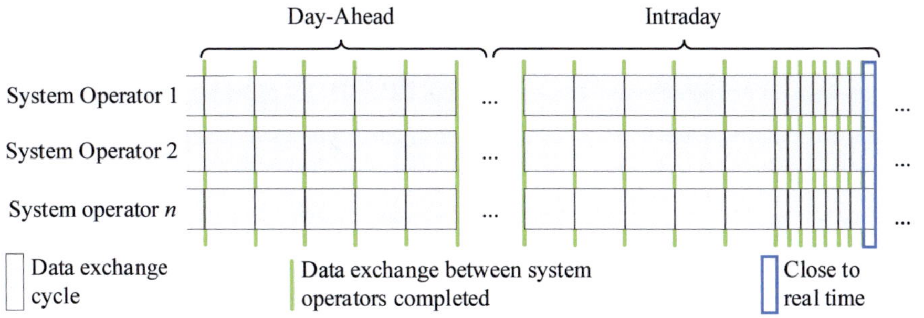

Fig. 2.4: Diagram illustrating TSO/DSO coordination (adapted from [110])

2.2.5 Data Modelling and Data Exchange

The increasing automation in the power system entails a higher volume of data processing in the control centre. Consequently, the universal deployment of ICT and the application of standardised data structures and protocols must be realised in order to cope with the increasing volumes of data [111]. To support the trend towards model driven architectures and their integration across system boundaries, in terms of data semantics and syntax, standardized data models have been developed [112]. The reference architecture for power system management and associated information exchange based on IEC TR 62357-1 [112] is shown in Fig. 2.5. It comprises the Common Information Model (CIM) based Common Grid Model Exchange Specification (CGMES), which has been developed for seamless integration and data exchange between several entities of the power system. The implementation process resulted in a unification of data structures and the extension of CIM by new classes to map components that were not yet standardised, which effectively improved the interoperability of devices and their CIM compliance. The CIM standard can be subdivided into the series IEC 61970 for basic requirements of energy management systems, IEC 61968 for distribution management and IEC 62325 for energy market communications, which consist of numerous parts that are constantly being further developed [113]. The standard series IEC 61970 [114] describes the interface for application programs in energy management systems (EMS-API) and comprises specifications of components in energy management or network control systems. Besides the definition of a naming convention and the data structure, it also aims towards a clear and consistent assignment of data. This allows to merge models of different detail levels, e.g., of neighbouring TSOs into one model. For EMS applications, the data model of the power system needs to be maintainable, scalable, interconnectable and exchangeable. A CIM/CGMES data set consists of several standard profiles, each containing a data set with different information objects. These profiles differentiate between equipment objects and container objects. Containers are objects such as voltage levels or switchgear, while equipment objects represent real equipment such as transformers or transmission lines. Thus, the CIM/GMES provides a modelling framework with suitable model classes, their attributes, and topological relations. The number of CGMES profiles will be extended in future by additional information models (e.g. availability planning, remedial action schemes, and system integrity protection) [115]. The latest edition of the CIM/CGMES is a set of the underlying standards summarized in IEC 61970-CGMES:2022 [114]. The CGMES support several EMS applications, such as [107]:

- Power flow and contingency analysis,
- Short circuit calculation,
- Capacity calculation, capacity allocation and congestion management, and
- Dynamic security assessment.

In addition to the data modelling standards, the architecture illustrated in Fig. 2.5 comprises different power system automation protocols, which are utilised for the communication between control centres, substations, and field devices. The IEC 60870-6 standard has been developed for vendor independent communication of data sets between EMS [116]. The IEC 61850 is usually used for intra substation communication and protection, but the enhancements and amendments include also other applications [117]. In comparison to the legacy telecontrol standards (e.g., 60870-5-101/-104 [118, 119] or DNP3 [120]), which exchange data via a strictly defined telegram structure, the IEC 61850 has semantic advantages in terms to model the power system devices ant their functions, which enable a higher degree of process automation. Although the legacy standards are quite powerful, they become obsolescent, where new power system infrastructure is installed. Therefore, vendors of RTUs will presumably discontinue the support of IEC 60870-5 and DNP3 at some point in the future. In countries where these protocols are still in use for historical and technical reasons, the interest in IEC 61850 is growing steadily and obsolescence of the legacy protocols is only a matter of time. A comparison of IEC 60870-5-101/-103/-104 and IEC 60870-6 with IEC 61850 has been published before in [121]. Specifications for secure exchange of information applicable for the aforementioned standards is given in IEC 62351 [122].

As illustrated in Fig. 2.5, a major part of power system automation systems is located at substations, which inherit a bunch of instrumentation to facilitate functions including monitoring, control and communication [123]. These functions comprise a wide range from switching a few feeders up to several cascading tasks, such as protection, automation, and complex control systems.

A generic substation configuration and the corresponding IEC 61850 data model is illustrated in Fig. 2.6. The switch bays contain circuit breakers and isolators, which connect the primary equipment such as transmission lines or transformers. Each feeder is also equipped with secondary equipment, namely voltage and current transducers for monitoring and protection purposes. Within a digitally enabled substation, every analogue value will be digitised at its interface. Thus, the state and behaviour of the power system and primary equipment are instantly available as digital data [124]. Aiming for automation and faster response to critical operating situations, a trend towards decentralisation of monitoring and control functions exists [125].

According to the needs of the system operator, the process information is aggregated at substation level to reduce the amount of incoming data. Thus, only relevant information accumulates in the main control centre EMS [126]. This includes process data, as well as further information from external data sources (e.g., weather services, neighbouring system operators or plant operators).

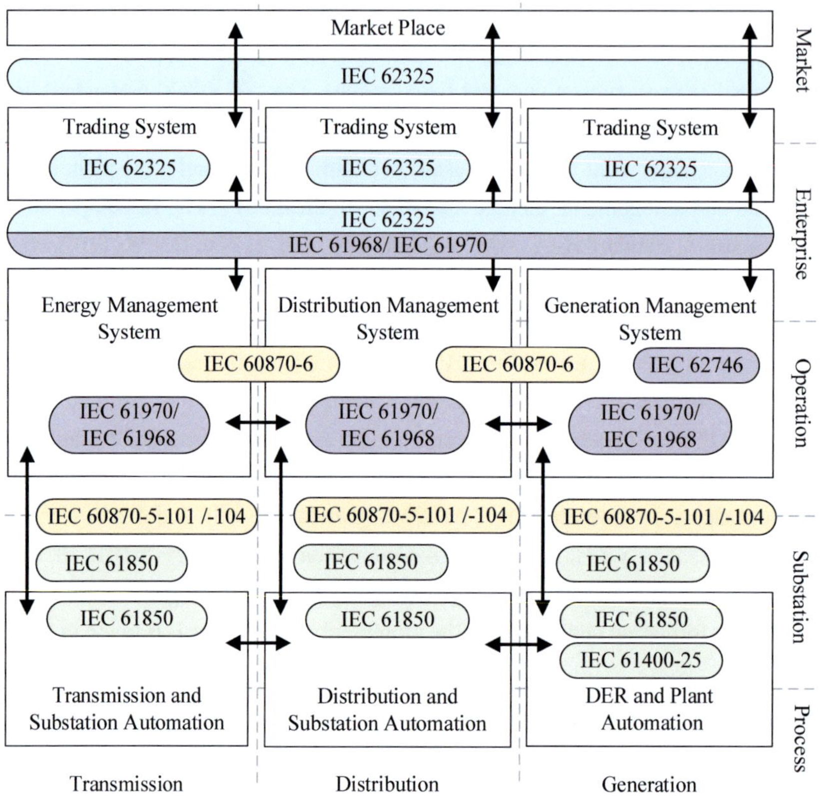

Fig. 2.5: Selected parts of the reference architecture for power system management and associated information exchange (adapted from IEC TR 62357-1 [112])

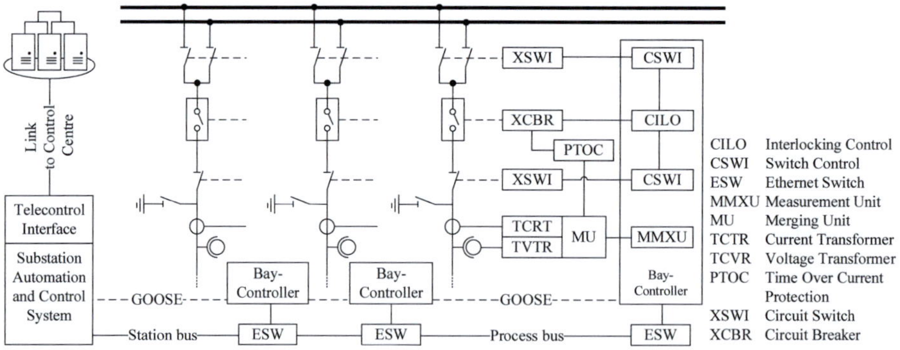

Fig. 2.6: Generic substation configuration scheme illustrating the IEC 61850 data model

2.3 Energy Management Systems

The system operation of transmission and distribution networks is carried out by means of EMS or distribution management systems (DMS) from a control room in the grid control centre. Thus, the EMS applied in the control room in the control centre constitutes the nerve centre of the power system [127], [128]. In large area power systems, tasks are commonly split to several control centres to supervise the responsibility area. Depending on the voltage level of the observed control area, different control centres exist with different tasks, degrees of observability and automation (main transmission control centre, sub-transmission control centre, distribution control centre, etc.). While transmission system operators obtain a high degree of automation and are usually well equipped with telemetry and telecontrol devices in their responsibility area, distribution system operators, do not automatically acquire measurement values from the overall area of responsibility. Especially in the medium and low voltage levels of distribution systems, observability and controllability is rarely given.

EMS in the control centre have evolved from hardwired analogue systems with manual controls into highly-integrated software suites [8, 15, 18, 127, 129]. They provide a wide range of functionalities customised to the individual needs of the system operator and can be divided into two functional groups [127, 129]. The primary functions comprise all aspects of information processing, i.e., handling of the communication via the SCADA system to monitor and control the remote electrical equipment. In addition to the basic SCADA functions, including the telecontrol links to remote electrical equipment (see also subsection 2.3.1), the EMS primary functions include a front-end system for communication, an information model management system. It contains a topological system model, the telemetry and telecontrol addresses, as well as relevant electrical model parameters. The EMS functionalities can be accessed via workstation clients by a suitable graphical user interface (GUI) and comprise gathered telemetry data. From this information, a state estimator (SE) creates a computable steady-state image of the power system. The secondary functions provide model-based statements and decision-support tools for the process of power system operation and operational planning and depend on the data provided by the SE. They are also referred to as the advanced decision and optimization functions (German: *Höhere Entscheidungs- und Optimierungsfunktionen* (HEO). Among other things, the HEO functionalities comprise contingency screening, security assessment, switching simulations and power flow optimisation. A schematic control centre EMS architecture is illustrated in Fig. 2.7.. Furthermore, a historical data archive and a non-repudiation able operational logbook to document all actions of the operators exist. Depending on the configuration, the process data is stored in the archive and often compressed for long-term archiving. Most EMS and HEO applications are interacting, and interdependencies exist as illustrated in Fig. 2.8.

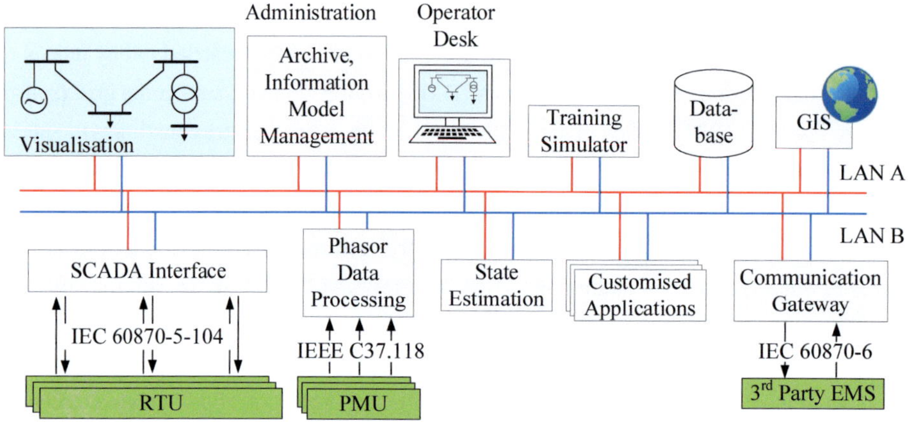

Fig. 2.7: Simplified EMS architecture (see also [130])

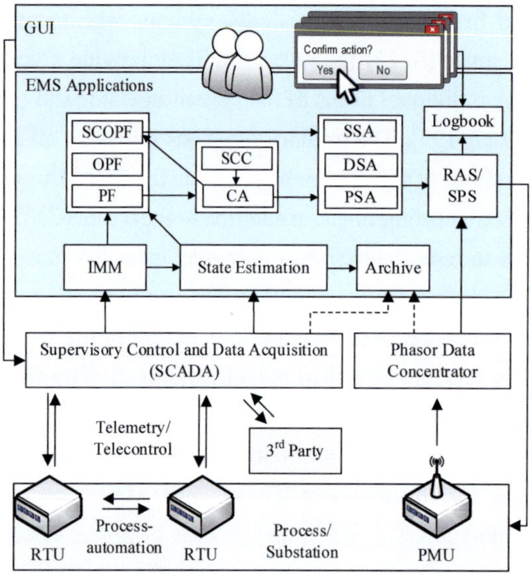

GUI: Graphical user interface
EMS: Energy management system
PF: Power flow
OPF: Optimal power flow
SCOPF: Security constrained OPF
IMM: Information model management
SCC: Short circuit calculation
CA: Contingency analysis
SSA: Static security assessment
DSA: Dynamic security assessment
PSA: Protection security assessment
RAS: Remedial action scheme
SPS: Special protection scheme
RTU: Remote terminal unit
PMU: Phasor measurement unit

Fig. 2.8: Simplified interdependencies of EMS calculation modules and applications

HEO applications require topological information, and operational degrees of freedom e.g., load and generation forecasts, generation costs, as well as generation reserves, to achieve the optimization targets. By defining an objective function either an optimal solution for active and reactive power flow (economic dispatch) or optimal reactive power flow (minimum losses) can be found applying optimal power flow [131], [132]. Common applications for protection comprise short circuit current calculation, and protection security assessment (PSA) to evaluate system protection settings, selectivity, and protection coordination.

2.3.1 Supervisory Control and Data Acquisition

The term Supervisory Control and Data Acquisition (SCADA) generally refers to the monitoring, control, and visualisation of a technical process by computer systems that communicate with each other via an ICT network. According to the standard for SCADA and automation systems IEEE C37.1 [133], a SCADA system as illustrated in Fig. 2.9 (adapted from [134]) comprises the following major components:

- *The master station,* in the control centre, where the operators monitor the system and take control, if necessary, via a *user interface*, which allows interaction with the power system,
- *Remote terminal units,* at the substations that communicate with the master station to enable data acquisition and handling control commands for the field devices,
- *A communication system*, which transmits telemetry and telecontrol data between the substations and the data aggregator in the control centre.

The main architecture of SCADA systems comprises two layers. While the client-layer allows human interaction with the supervised system, the supervision itself is carried out by the server-layer. The operator terminals (UI-clients) can be accessed locally in the control room or remotely by authorised personnel. The UI clients allow access to different functions depending on the role and access rights of the employee. A modern SCADA system consist of the following components:

- A graphical representation of the system variables and switching state,
- Configurable information sheets and diagrams,
- A real-time information display and a study mode that can be clearly distinguished,
- An alarm list and access to the operator logbook (irrefutable documentation of all operational actions taken),
- Access to ICT and subsystem status information and maintenance functions,
- An operating environment with access to telecontrol functions.

SCADA messages include status information or status changes, such as breaker and switch positions, measurement values, setpoint commands, switching commands, control commands and time synchronization signals. The data transmission is realised by so-called Remote Terminal Units (RTU) in the substations by short telegram messages (see also subsection 2.2.5).

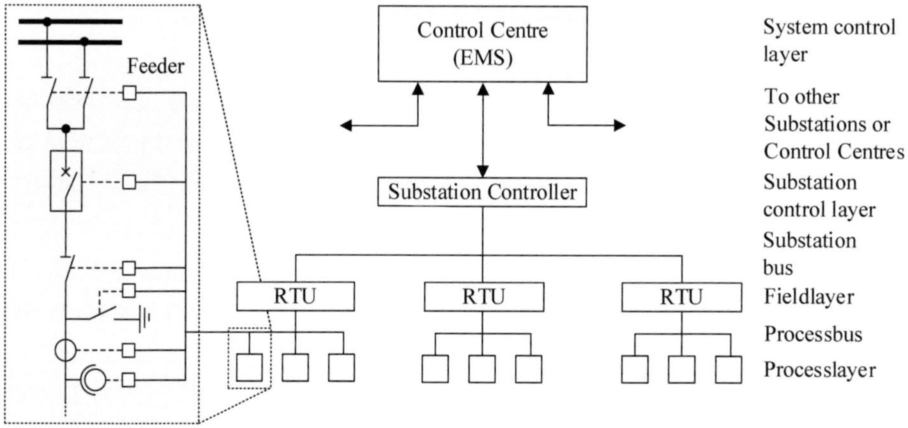

Fig. 2.9: Schematic representation of a SCADA system (adapted from [134])

These arrive cyclically or spontaneous and contain the information itself, the cause of transmission and other relevant meta-data. The report rate depends on the device settings, respectively the polling cycle. SCADA systems usually obtain new measurements sets every 1–10 seconds [128].

In transmission systems, the communication is almost entirely realised with fibre optic cables, while in distribution networks, due to economic reasons ripple control or other common telecommunication standards (LTE, GPRS, ...) are often in use to a large extent.

The process information is made available by field devices in the substation. In some cases, the raw data is re-processed in the substation and aggregated within the transmitted telegrams. Messages, measurement, and count values are timestamped and archived in the SCADA database. Besides measurement and count values, average and extreme values are recorded continuously within time resolutions of up to one second. This enables the creation of generation and load profiles, which allow extrapolation and forecasting. Alarms and events (e.g. malfunctions or limit violations), have to be acknowledged to ensure that they are actively perceived by the operator [135]. The SCADA system creates a process image, which is stored in a database on a regular basis and utilised by the secondary EMS functions (see subsection 2.3). It contains topology and switching state information, as well as state variables received by RTUs. A simplified power system representation is used to present the current operational state of the system to the operator. In software-based SCADA systems, switching procedures and other telecontrol actions are enabled within the context menu of symbolic representation of the respective asset. In case the telemetry link is not available or the telecontrolled devices are malfunctioning, the last valid data set will be used to maintain a valid process image. The basic performance requirements for data acquisition and control in automation systems are given in Tab. 2.3

Tab. 2.3: Basic requirements of SCADA systems (derived from [133])

Performance	Data acquisition and control
High availability and reliability	Task conform measurement data resolution
Low latency and time skew	High resolution and accuracy
Data security (access rights management)	Timeliness and completeness
System compatibility (open architecture and system interface standards)	Remote control and access to relevant system variables
Redundancy	Reliability

2.3.2 State Estimation

Due to missing or bad measurement data or (temporary) communication failure, full observability of the power system is not always given. The telemetered data naturally contains measurement inaccuracies, conversion errors, time skew or communication noise. A state estimation (SE) is therefore applied to compensate sources of uncertainty, to consider communication delays, measurement noise, missing or false data. It creates a consistent and reliable data set of approximated values. Therefore, redundant measurements are needed, to enable bad data detection, to correct erroneous data and topological errors. The SE module is often called the corner stone of the control centre EMS, since the most secondary EMS applications, such as economic dispatch or security assessment are built upon it. It uses available telemetered data from the SCADA system (injected active and reactive power at the buses, branch flows, and bus voltages), to compute the state of the power system, which satisfy the algebraic power flow equations consistently and with a high statistical probability. Due to the formulation of the SE problem as an overdetermined system of non-linear equations, the solution is calculated numerically within an iterative procedure. The concept of the power system SE has been described by Schweppe [136–138], followed by several proposals for improvements and practical considerations [139–142]. The traditional SE methodology, which is utilised worldwide the electric power industry, applies the method of weighted least squares to approximate a steady state representation of the power system [143, 144]. Since the traditional least squares SE algorithm is not very robust against measurement outliers or large deviations [145], bad data detection routines are required. A prerequisite for an effective estimation of the system state is a properly maintained database of electrical equipment characteristics and type related parameters to set up an accurate network model. Furthermore, a covariance matrix containing the approximated measurement error variances is required. The SE is executed cyclically or manually on demand (e.g., after a SCADA general interrogation) to maintain observability. It is illustrated in Fig. 2.10.

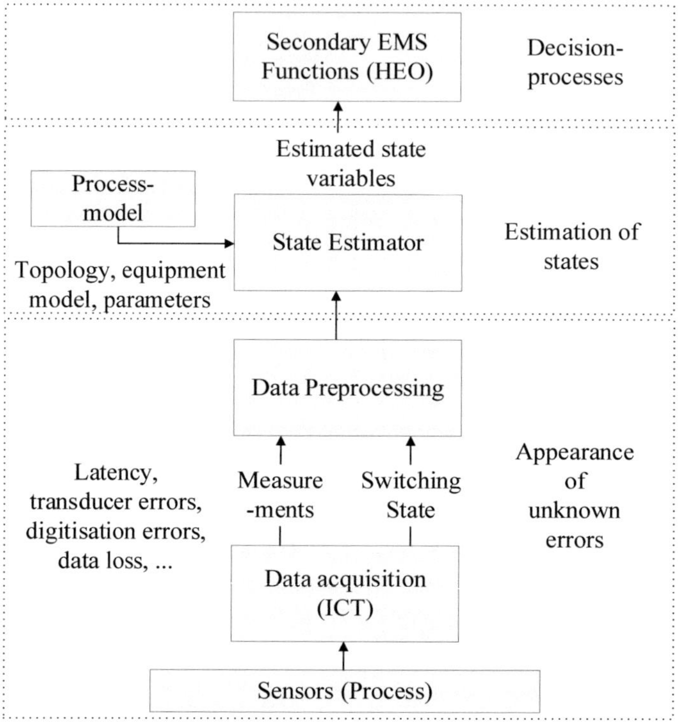

Fig. 2.10: Schematic representation of the power system state estimation process

With rise of the synchrophasor technology and PMU deployment, linear and hybrid state estimators evolved [146–149]. If a certain degree of observability by PMUs is given, the utilisation of complex bus and branch currents instead of active and reactive branch flows, enable fast linear SE techniques [150]. The calculation time of large SE problems can be reduced by the application of hierarchical or distributed SE schemes [151–153]. A pre-processing SE method at substation level has been described in [154].

The term dynamic state estimation (DSE) has been introduced by the power systems community to express the ability of the respective estimation techniques to identify parameters of the nonlinear dynamic models and to predict dynamic state variables of time-domain models, such as power plant models, inverters, or dynamic loads. The dynamic state variables are referring to the set of variables that are required to determine the condition of a dynamic system at a certain time (e.g. machine rotor angle or speed) [155]. Several methods and algorithmic solutions have been developed in the past decades to estimate the state of a dynamic system when the states are not obtainable by direct measurements. Thereby, the term observer is often used synonymously to an estimator of unknown or inaccessible states (e.g., in [156]). An observer is applied for control purposes to estimate state variables, when the full state vector (i.e., observability) cannot be determined by measurements, or the measured outputs are dependent of unknown or undetermined states. When applied, the objective of a state observer is to minimize the deviation between an estimated and a measured signal, by applying a deterministic model. If the error between both values is small, the probability is high, that the estimated value is close to the real value, assuming that the applied model is correct. Due to their deterministic approach, observers are sensitive to noise, thus low noise sensors with a high-resolution are required for good performance [157]. To estimate states subjected to stochastic noise or disturbed signals, the Kalman Filter [158–160] and its derivatives [161], [162] have been developed. Another promising state estimation technique is the moving horizon estimator. It considers a trajectory of states instead of a single time step. This positively impacts accuracy of the results, at the cost of computation time [163], [164]. The current, and potential future of DSE applications for power system modelling, monitoring and control, as well as their critical importance for the development of future EMS has been outlined in [165, 166] and summarised in [167].

Although hybrid, linear and dynamic state estimation techniques are often discussed in research, utilisation within operational EMS environments are still uncommon, due to their strong dependence on reliable ICT [165]. Algorithms, improving DSE speed and robustness [168–171], and for advanced applications, such as model predictive control [172, 173], online DSA [174, 175], as well as for wide area control [176, 177] have been proposed in the literature. However, the next-generation EMS itself and a future proof architecture applying DSE for system operation has not yet been addressed in detail. The main difference between SSE and DSE are their field of application, modelling, and data requirement. While SSE is applied for dispatching and monitoring of bus voltage and branch flow limit violations, DSE is applied within the context of DSA and time-domain model calibration. Most DSE techniques used for power system applications apply the class of Kalman filters. These often depend on data streams provided by PMU based WAMS [178–184].

A general overview of DSE algorithms and possible applications is given in [185], [186], whereas available algorithms for DSE in power systems have been evaluated in [187]. The approaches for state estimation discussed beforehand are compared in Tab. 2.4.

Tab. 2.4: Comparison of state estimation approaches

	Classic SE	Hybrid SE	Linear SE	Dynamic SE
Data source	SCADA	SCADA and PMU	PMU	PMU
Model basis	Algebraic steady state power flow	Algebraic steady state power flow	Algebraic dynamic power flow	Differential-algebraic equation-based time-domain model
Input data	Bus voltage magnitudes: $U_{i,j}$; Bus injections: $P_{i,j}$, $Q_{i,j}$, I_{ij}; Branch admittance: $\underline{Y} = G_{ij},+jB_{ij}$	Bus voltage magnitudes $U_{i,j}$; selected $\underline{U}(t)$; Bus injections $P_{i,j}$ $Q_{i,j}$; Branch flows P_{ij}, Q_{ij}, I_{ij}; Branch admittance $\underline{Y} = G_{ij},+jB_{ij}$	Complex bus voltages $\underline{U}_{i,j}(t)$; Branch admittance $\underline{Y} = G_{ij},+jB_{ij}$	Complex bus voltages $\underline{U}_{i,j}(t)$;$_{i,j}$; Branch currents $\underline{I}(t)$; Parameter set for equipment and controller models
Resulting state variables	$\underline{U}_{i,j}$, branch flows, transformer tap positions, switching state (corrected topology)	$\underline{U}_{i,j}$, branch flows	$\underline{U}_{i,j}$, branch flows, topology	Algebraic states: $[z_1,...,z_n]$, differential states: $[x_1,...,x_n]$, and state derivatives: $[\dot{x}_1,...,\dot{x}_n]$
Complexity and computing time	Medium	Medium	Low	High
EMS implementation	State of the art	Some implementations	Applied by WAMS	Scarce implementations, experimental state

2.3.3 Security Assessment

The term *security* in the context of power systems refers to *"the ability of the power system to withstand sudden disturbances"* without interruption to customer service [188, 189]. It is closely related to the term *stability,* which can be defined as *"the ability of an electric power system, for a given initial operating condition, to regain a state of operating equilibrium after being subjected to a physical disturbance"* [188, 189]. Both terms, *security* and *stability* are aspects of *reliability*, which refers to *"the probability of satisfactory operation over the long run"* and *"denotes the ability to supply adequate electric service on a nearly continuous basis"* [188, 189]. Following this definition, security assessment tools evaluate the ability of the power system to withstand perturbations and to sustain the system reliably under adverse conditions. Security assessment can be subdivided into three levels [190]:

- *Security monitoring*, to check if the operating conditions are satisfied,
- *Security analysis*, to prove that the system is able to withstand specified disturbances, and
- *Security margin determination*, to evaluate the level of security and resilience for a given operating condition.

To analyse power system security, two possible techniques are applicable, depending on the requirements of the investigated criteria [190]:

- Static security assessment (SSA)
- Dynamic security assessment (DSA)

SSA tools evaluate post-disturbance conditions, based on power flow calculations. Assuming that the dynamic transition from pre-to post-contingency does not cause instabilities, they are applied to verify that no equipment ratings and voltage constraints are violated. DSA methodologies evaluate the stability of the transient transitions between pre- and post-disturbance states, to ensure that the system stabilises after the disturbance. DSA tools consider the damping of transients, their amplitude and the impact on the quality of service. Both techniques are briefly discussed in the following subsections.

Static Security Assessment

Static security assessment (SSA) is based on power flow studies to investigate the violation of operational limits in steady state. Since all credible contingencies are screened, beginning from a base scenario derived from the SE module, SSA comprises a contingency analysis (CA). Applications for CA are screening the actual and prospective topological switching state of the grid for congestions or user specified stability in terms of *N*-1 and sometimes *N*-2 security. In case that a contingency is leading to unacceptable conditions, countermeasures need to be identified. Conditional variables of interest are exceeding voltage limits, ampacity limits, thermal line loadings, or generator capability violations. A CA is done for each single component with the actual power flow data from the SCADA system. The CA alters the topological configuration and evaluates the loading to investigate the possible violation of operational limits for neighbouring equipment within the observed area. To reduce the time required for the identification of contingencies, a prioritised list of the most relevant equipment can be compiled. Alternatively, algorithms that handle the CA by approximation e.g., DC power flow in a first run, and hand over those cases to the conventional CA algorithm for detailed analysis, are applied [129]. Another approach is the application of power flow algorithms, which apply state changes directly in the inverted nodal matrix of the network model, without requiring a new inversion [191, 192]. Lately, techniques applying machine learning have gained interest in research. Investigations have shown that they promise to address technical shortcomings of the traditional approaches in terms of computational speed and accuracy, especially when applied online [193], [194].

In all techniques, the *N*-1 criterion, is considered. Since the loss of equipment leads to significantly higher loading of neighbouring equipment, the *N*-1 criterion guarantees a certain degree of resilience (see also subsection 2.2.2). It guarantees that the loss or outage of an important system component does not lead to constraint violations and normal system operation is still possible.

Dynamic Security Assessment

The conventional concept of steady state security verification, cannot sufficiently guarantee the system integrity in all situations [71]. Therefore, it is mandatory for TSOs to assess dynamic stability. The ENTO-E demands that "*all TSOs of each synchronous area shall coordinate the dynamic stability assessments, which shall cover all or parts of the synchronous area* [38]."

The terms online and offline DSA are applied to distinguish between DSA applied during power system operation and power system planning phase. To identify stability limits, dynamic simulation models from system planning department are applied during the operational planning phase (see subsection 2.2.1), to reproduce system dynamics and to estimate security margins for system operation. These offline determined limits inherit several uncertainties, i.e., the finally applied limits in the real time environment are not considering all possible operational conditions [195]. This may result in situations where information about system security is unavailable. To eliminate those uncertainties, the need for online DSA has been recognized since the 1970's [40]. Early DSA systems have been deployed in the late 1980's [196].

To assess dynamic security online, the time-domain models available from the planning department are coupled to the real-time process information. This enables to investigate the actual operating point or a region of interest around this state, and addresses also unexpected operational conditions that have not been investigated before in the process of power system planning.

An offline DSA study may address long term grid planning, medium term offline DSA may help to avoid potentially critical states and identify preventive or remedial measures, while online DSA can support decisions during real-time operation [71]. The ENTSO-E recommends a continuous validation and fine tuning of system models (e.g. by WAMS) to avoid transient conditions within the numerical simulation, where electrical variables could violate validity boundaries of the model, leading to invalid results and in worst case wrong decisions [71]. The same applies to correct dynamic model initialisation from process information since incorrect initialization of state variables leads to wrong simulation results and conclusions.

The evaluation of contingencies and their impact on system security considers several criteria derived from the classical power system stability definitions. These DSA criteria include rotor angle stability, voltage stability and frequency stability, as well as the damping of oscillatory modes of machine and controller interactions with the interconnected transmission system. Furthermore, compliance to boundary definitions from grid codes or other specified thresholds can be evaluated [197, 198]. The assessment of the mixed stability criteria is done by a composite index which represents weighted stability criteria according to the system operator needs. The applicable criteria are described in the literature [199, 200]. To visualise the stability index ranking system, a signal light approach can be used [199]. The simplified flow chart of a DSA assessment tool and a visualisation cockpit as applied in control centres is shown in Fig. 2.11.

DSA systems are not state of the art in EMS. A lack of accurate dynamic models, missing guidelines how to use DSA information for real-time system operation, unfamiliarity of control room operators about power system dynamics, and missing concepts how to apply dynamic stability limits are the main hindering aspects for a wider adoption by TSOs [71]. Recent research on DSA methods and industrial implementations focus on predictive and automated online DSA for application in operator decision support systems e.g. in [195, 201], or in [125, 202]. Approaches to apply DSE techniques for direct initialisation of the dynamic model equations from the latest estimated dynamic state have also been investigated [167]. A review on DSA tools and techniques is given in [190], and [203].

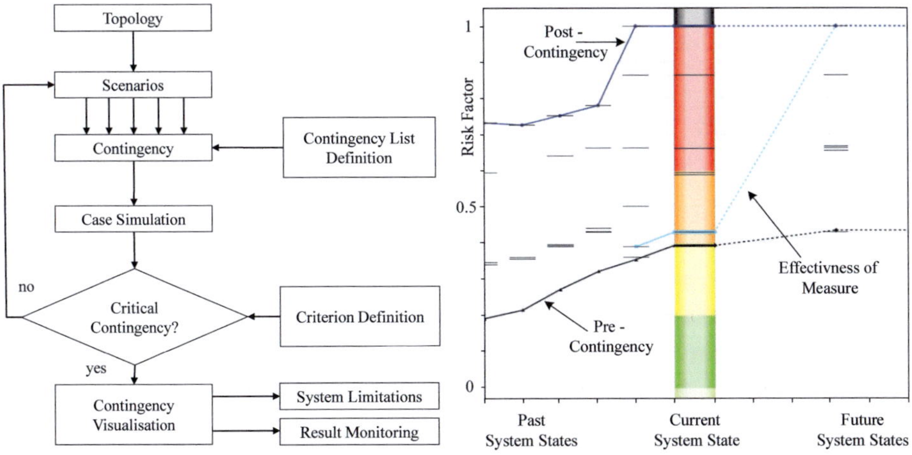

Fig. 2.11: Structure of SIGUARD® DSA assessment and visualisation cockpit [72, 204]

2.4 Wide Area Monitoring, Protection and Control

Although wide area monitoring systems (WAMS) can apply all available metering devices which are able to communicate, it has been the massive deployment of PMUs, which led to worldwide WAMS application. These provide dynamic observability and operator decision support within control rooms [205]. In comparison to traditional SCADA systems, the main advantage of WAMS is the application of time synchronized PMU samples, which allow efficient handling and visualisation. As described in subsection 2.4.1, PMUs enable a centralised monitoring of system dynamics in the control centre EMS. Besides WAMS (see subsection 2.4.2) also wide area protection systems (WAPS) and wide area control systems (WACS) have been developed. These are described in subsection 2.4.3. To summarise wide area monitoring, protection and control applications, the acronym WAMPAC is sometimes used in the literature.

2.4.1 Phasor Measurement Units

A phasor represents a periodic signal with a fixed frequency by a complex equivalent of the sinusoidal wave entity. The principle of the phasor, as a complex quantity representing a sinusoidal AC signal, was already described in 1893 by Charles Proteus Steinmetz [206]. In the complex plane, a phasor is represented by the vector, which is determined by the amplitude $\hat{x}$ and the associated angle φ. As a phasor is only valid for a fixed frequency, it is displayed independent of this frequency (2.1).

$$x(t) = \hat{x}\cos(\omega t + \varphi) \tag{2.1}$$

According to equation (2.2) the active power flow between two terminals connected through a transmission line is almost proportional to their voltage angle difference. Hence, the measurement of the phase angle differences across transmission systems offers advantages for power system observability.

$$P_{ij} = \frac{(U_i U_j)}{X_{ij}} \sin(\theta_i - \theta_j) \tag{2.2}$$

The required time synchronised phasor measurements are derived by PMUs. These devices analyse the continuous analogue signals of AC voltage or current derived from current and voltage transducers, and translate them into digital samples. By examining the parameters of equation (2.1) by a discrete Fourier transformation, PMUs derive magnitude, phase, sequence, frequency, and rate of change frequency (ROCOF), of the alternating signals given at its input terminal. The deployment of PMUs is dominant in transmission systems, but is also gradually rising in distribution grids to raise observability [207, 208].

Besides the measurement samples, a PMU frame comprises a frame header. It contains descriptive information about the PMU configuration, such as the cyclic redundancy check (CRC) codes, and a quantifier for the quality of the measurement sample [209] (see also Fig. 2.12). The time signal is given by a precise time source, such as a Global Positioning System (GPS), which enables synchronisation of the measurements of several PMUs.

With at least two PMUs and an accurate GPS time reference with a resolution of 0.2 – 0.5 μs [210], the angle difference between two locations becomes measurable. Due to their synchronized time tag, phasor values obtained by PMUs are often also referred to as *synchrophasors*. As illustrated in Fig. 2.12, the measurement samples are digitised and packed into frames for transmission. A set of synchronised phasor measurements, the frequency, and the ROCOF that correspond to the same time stamp, is called a data frame [211]. The process of the creation of a PMU frame is illustrated in Fig. 2.12.

The IEEE standard C37.118 [209, 211], which is meanwhile partly superseded by the IEC/IEEE 60255-118-1 [212], is explicitly used to stream data from PMUs to the phasor data concentrator (PDC). As illustrated in Fig. 2.13, the data stream is forwarded to a phasor data processor (PDP) in the control centre for analysis and visualisation [209]. A visual summary of PMU related standards (e.g., PDC requirements [213]) is shown in Fig. 2.13.

The accuracy of PMU samples strongly depends on the instrument transformers and transducers at the measurement location. Since it is uncommon to characterise the nonlinear behaviour of transducers in industrial practise, the compensation of measurement errors only possible to a limited extend [214]. The accuracy of PMU devices are divided into classes P and M. The class P is intended for protection applications, which require fast response and mandate no explicit filtering. The class M is intended for measurement applications that could be adversely effected by aliased or biased signals [211].

Most PMUs support reporting rates, from 10 samples/second up to 120 samples/second for 60 Hz power systems [211]. The requirement for the time tag accuracy is ±1 μs [211], which corresponds to an angle error of 0.018 degrees for a 50 Hz system. The PMU error is expressed by the total vector error (TVE) and is calculated according to appendix A.4. To comply to the IEEE standard C37.242-2013 [215], systematic errors of up to one percent in the magnitude and one degree in the phasor angle may be introduced by the remaining measurement chain. Modern PMUs can process measurements received from an IEC 61850 device via the substation process bus, skipping the analogue acquisition from transducers [216].

A profound description of the technical development and applications of PMUs is given in [217]. In transmission systems the observability level is generally high due to an almost full coverage of RTUs in substations and a rising number of installed PMUs [218, 219].

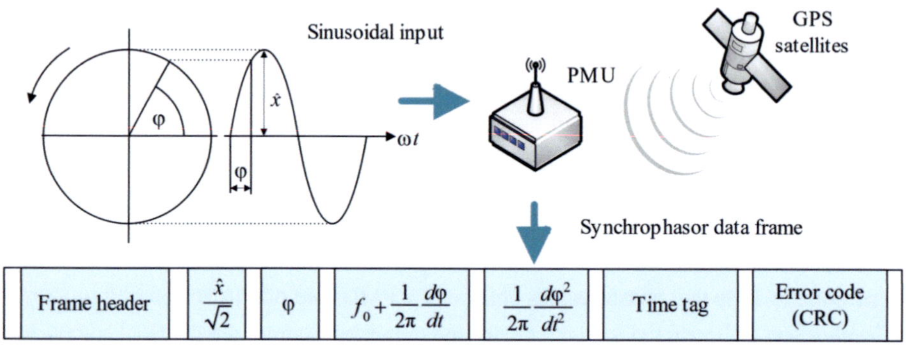

Fig. 2.12: Scheme of the creation of synchrophasor data frames by a PMU (adapted from [220])

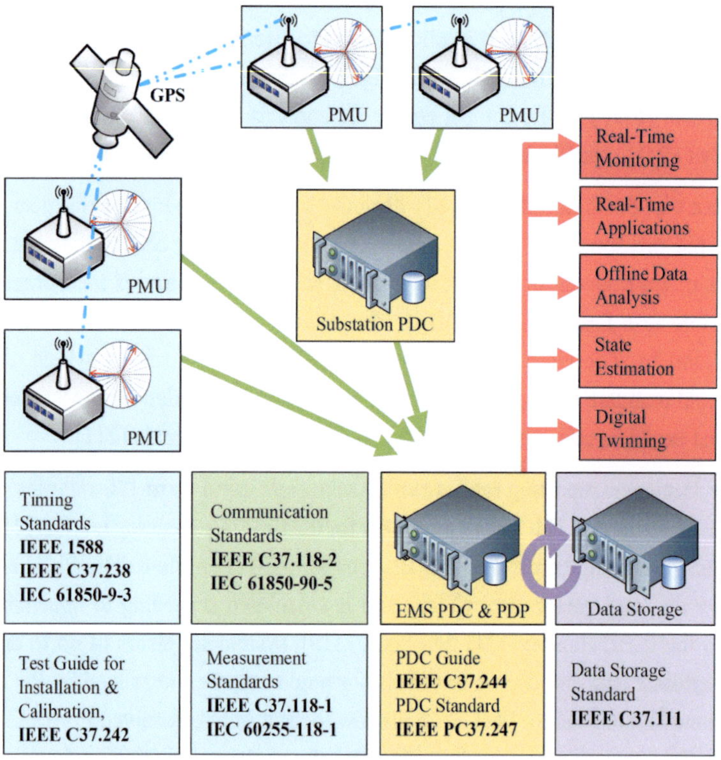

Fig. 2.13: Visual summary of PMU related standards (adapted from [221] and previously published in [69])

2.4.2 Wide Area Monitoring Systems

Wide area monitoring systems (WAMS) are designed to provide a continuous dynamic power system state monitoring. In transmission systems, WAMS are mainly PMU based, and add an additional observability layer running independently from the SCADA system. It enables observation and analysis of dynamic events, even if the SCADA system is dysfunctional [19]. Although it is possible to supplement PMU data stream based WAMS by SCADA measurements, it can lead to a heterogeneous data foundation, which raises problems of synchronicity and alignment for data processing [222]. State-of-the-art applications and implementations of WAMS into control room environments are described in [216], [222], [223], [224].

Although WAMS are deployed worldwide for monitoring at control centres [225], some TSOs do not include WAMS into operation until clear procedures are defined which help operators to relate WAMS information to control actions to be taken [216]. Thus, obstacles for a wider adoption of synchrophasor technology in operational control rooms are mainly driven by the challenge to purposefully apply the new type of information provided by WAMS. The integration of WAMS information, measurements and alarms into SCADA systems are common applications and represent a useful intermediate solution for supplementing SCADA information [216]. Therefore, strategically important PMUs feed WAMS across areas of responsibility in some cases [226].

WAMS functionalities can be separated into online or offline applications. Online applications are applied for real-time power system operation. Offline applications of WAMS can be subdivided into forensic or post-mortem analyses of historical events, support in system planning (e.g., evaluation of frequency responses) and validation of system models. An overview of WAMS offline and online applications is summarised in Tab. 2.5.

2.4.3 Wide Area Protection and Control

The information derived by WAMS for feedback control and system protection is applied in a few control room implementations worldwide. An efficient application of wide area control systems (WACS) into power system operation, requires flexible and controllable utilities, such as phase shifting transformers (PSTs), FACTS devices or HVDC systems. Experiences with WACS based closed loop control has been made with wide area damping control [227], [228], power flow control by FACTS, coordinated PSTs and HVDC links [216]. Emerging applications based on phase angle differences derived from PMUs, allow automated remedial action schemes [91, 96] and advanced emergency protection schemes (see subsection 2.2.2) [216].

Wide area protection schemes (WAPS) using WAMS as an input are also well established. They are rather cheap and easy to install, but a major drawback is, that they inherit the risk of a large-scale blackout in case the system fails and no fallback is available [229]. Thus, automated WAPS will be applied in practice only if the communication system is reliable and secured, and the applied rules are transparent and simple [230]. Systems considering all aspects of wide area monitoring, protection and control are referred to as WAMPAC [231]. Several possible applications of WAMPAC systems are summarised in Tab. 2.5.

Tab. 2.5: Summary of possible online and offline applications of WAMPAC systems

Application		**Description**
Online	Data visualisation	Real-time visualization of PMU data streams [216, 232]
	Equipment monitoring	Monitoring of operational limits and alarming in case of limit violation, reducing uncertainty in security limits [216, 226, 233]
	Oscillation monitoring and damping	Analysis of local and inter-area oscillations [226] as well as damping levels [227, 232, 233], and resonance warning [216]
	Improvement of state estimation results	Enabling hybrid and linear state estimation approaches [232, 233]
	Frequency stability monitoring	Real time frequency and ROCOF assessment [231, 234]
	Voltage stability monitoring	Assessment of voltage transients, static and dynamic voltage stability indicators, and the distance to stability boundaries [232]
	Voltage angle monitoring	Estimation of the transmission capacity reserve by monitoring of the voltage angle difference between two locations [231]
	Disturbance classification	Approximation of the location of a disturbance and recognition of characteristic pattern to classify the disturbance type [224,
	Island detection	Assessment of frequency and phase angle values to indicate system splits or separation [216, 235]
	Inertia monitoring	Approximate estimation of inertia within the synchronous area [236], [216]
	Generator monitoring	Monitoring of power plants and synchronous generators [232], angular stability and pole slip recognition [237]
	Parameter estimation	Identification of transmission line and other power system parameters from the measurements [233], [236]
	System restoration	Adaptive load shedding, island management and resynchronisation [226], [238]
	System protection schemes	Automated remedial measures for system protection, (e.g. automatic switching, transformer or shunt tapping, HVDC setpoint adaption, generation curtailment, or protection setting adjust-
Offline	Retrospective analysis	Offline disturbance analysis, post-mortem or forensic analyses of relevant events [216, 226]
	System model validation	Validation and benchmarking of dynamic system models for operational and long term planning purpose [216, 233, 237]
	KPI development	Statistical assessment of WAMS data together with SCADA data to develop and monitor KPIs for long-term improvements [24]

2.5 Operator Assistance Systems

Outdated or inadequate monitoring systems combined with slow response or human error have been identified as critical factors, that often contribute to cascading failures [239]. In addition, operators must take a larger number of quick decisions in daily operation [34]. Thus, the complexity of the power system has reached a level that requires extensive automation functions in system operation and management [25].

To assist humans in executing real-time process control decisions in complex systems, expert systems (ES) and decision support systems (DSS) have been introduced to support real-time process control and decision making in complex system operations [240]. Some are matured parts of power system management [241–243]. Early industrial applications of power system operator DSS have been introduced as ES in the 80s [244]. These have been defined as programs, which comprise knowledge concerning a problem encoded into software [17]. Their main tasks were alarm handling, event monitoring and analysis, system restoration support, fault localisation, and basic remedial control functions [242, 245]. However, the most of the early EMS decision support tools were adopted from the planning department, and their real-time control capability was limited [243].

Strengths and weaknesses of ES when compared to human involvement have been discussed in [246]. As expected, humans excel at creative tasks, whereas ES are better at repetitive, predictable tasks using pre-programmed rules and heuristics. In [241] a few minor conceptual differences between ES and DSS are pointed out, but most differences lie in the nomenclature. The two terms are often used interchangeably, but the DSS is used more frequently in recent literature. Contrary to static DSS implementations, machine learning (ML) based approaches enable computers to directly learn from data. Therefore, ML techniques are considered as the enabling technology for advanced DSS, which are also coined as assistant systems (AS). AS for deployment in power system control room environments are currently investigated, but are not yet state of the art. The general concept of AS has been previously discussed in [41].

An AS displays the actual state of the supervised process and indicates changes, where intervention is needed. To decide whether a situation is normal or different from normal state, a set of pre-defined context variables is required. In case a context variable is met within certain circumstances, the AS can gain the attention of the operator to react or to follow the decision options given. Context variables within a control room EMS can be alarms (e.g., limit violations, fault events, unexpected spontaneous changes, error status indicators) with a certain class of priority or defined acknowledgement rules. The rule set defines a context model, which is necessary to decide whether and how urgent an appropriate action is required. Following the general definition for AS in [247], they are *„interactive, user adaptive problem solving aids that understand what they do, accept input in terms of goals [...], or deduce such goals and [...] aim at solving them independently".*

Recent developments focus on semi-automated assistance systems, to tackle the challenges of a rising system complexity. Especially the need to handle a higher number of events in a limited time motivate to implement autonomous systems, which can take independent decisions and control measures automatically [22], [248]. However, due to technological and regulatory constraints, the concept points to a future that lies far ahead. A generalised view of tasks and responsibilities of the operator and the supporting functions of the assistant system are given in Tab. 2.6.

Assistance system solutions for power system operation have been introduced recently by several vendors e.g., by PSI as tools for security assessment and system optimization (SASO) [249, 250], and by Siemens as part of the SIGUARD® suite [251]. A newer system, namely "*Apogee*" [248, 252] has been developed to perform security analysis on forecasted near-future system conditions through data-analytics and modelling tools [50]. The systems have in common, that they address the needs of transmission system operators, and take available EMS information into account. Since the power system stability imposes constraints for ensuring system security, it is becoming a crucial issue for planning and operation procedures. Consequently, the time frame of dynamic system studies comprises long term planning studies as well as C2RT scenarios [71] (see also Fig. 2.1 in subsection 2.2.1).

Tab. 2.6: The human role supported by assistant systems in control rooms [41]

	Assistant System	Human Operator
Tasks	Monitoring	Understanding
	Diagnostics	Learning
	Prognostics	Reasoning
	Documentation and highlighting	Acknowledgment
	Decision support	Take actions

Operator AS address the timeframe between C2RT and real time. This implies that the solutions provided by the operator assistant system have to be readily evaluated within a specific timeslot (today: 15 minutes or faster according to actual scheduling and market processes). Thus, when schedules in system operation become tighter, the performance requirements also rise accordingly. While an AS gathers and aggregates data to information, the reasoning and decision making itself remains the responsibility of the operator (see Tab. 2.6). It has been discovered, that a high degree of computer-based automation aimed at relieving stress can also be contrary to its purpose [35]. Therefore AS should display information only when relevant to the operator to reduce negative effects on human perception and responsiveness [50].

Although increasing the level of automation has many benefits, however highly automated systems can leave operators somewhat detached from the observed system and they may end up in a state of low awareness of its actual condition [25]. In the context of automation, this is usually called the "*out-of-the-loop*" syndrome [36]. To address this issue, the proposed EMS and its underlying assistance functions can help to avoid this low awareness state. The assistance system works as a filter between the fast-changing system state and the state presented to the human operator. It enables human operators to perceive and interpret changes correctly and to react efficiently and quickly. This allows to improve the operational decision-making process by mitigating the risk of information overflow and keeps the human operator "*in the loop*".

The appropriate level of automation cannot be conclusively determined, but it is to be expected that assistance systems will be used in future EMSs. However, the final remaining task for control room operators in a highly automated system will be to define the rule set (context variables), and to monitor the correct operation of the automation system, and if not to take over himself. Thus, automation leaves at least responsibilities for abnormal conditions in the hand of human operators [35]. An exact model representation of the system and its behaviour in response to changes in the system state is required in order to create flexible and convenient decision support and assistant systems. Hence, the DT is the key concept, as it provides the necessary data basis for integrating novel assistance functions into the next generation of EMSs.

2.6 Parameter Estimation

Simulation tools for operational planning studies and real-time operation depend on the exact knowledge of the power system model parameters. These are typically calculated according to the geometrical information or estimated through measurements [214]. During operation, saturation effects, temperature variations, weather conditions and aging effects influence these parameters. To maintain accurate models, and to adapt those during operation to changing environmental and system conditions, parameter estimation techniques are required. Several techniques for transmission line parameter estimation exist, applying PMU and SCADA data (e.g., [253], [254], [255], [256], [257]). To estimate the full nodal admittance matrix $\boldsymbol{Y}_{\mathrm{bus}}$ the weighted least squares and the extended Kalman filter (EKF) have been applied in [258]. The authors find that the insufficient accuracy of the sensor data is a limiting factor to create an adaptive model under field conditions. For simulation studies in the time-domain, such as DSA, additional parameter sets are needed to model the behaviour of rotating machinery, converters, and their respective controllers. *Autoregressive moving average* (ARMA) model based parameter estimation techniques have been applied to estimate power plant parameters [259], or the inertia within the synchronous area [236]. DSE techniques (see subsection 2.3.2) are also applicable for wide area control and stability monitoring [166], as well as for system model validation and dynamic model parameter calibration. The Kalman Filter (KF) and its derivatives, such as the EKF and the unscented Kalman filter (UKF) or the cubature Kalman filter (CKF), have also been applied by researchers successfully [260, 261]. Recently, also moving horizon estimators (MHE) found application in the power system domain [262–264]. However, the scope of MHE application was not to determine model parameters, but to support conventional power system state estimation. Although available, it can be concluded, that parameter estimation techniques are not yet applied in EMS to derive model parameters from online measurements. Especially techniques for model parameter identification of controllers under varying system conditions is of interest for EMS environments and in scope of recent developments [237].

2.7 Conclusion and Gap Analysis

The analysis of the state-of-the-art power system monitoring, and control systems and EMS functionalities unveils, that modern control centres inherit complex hardware and software interactions over dedicated ICT with strong security, reliability, and accuracy requirements. Today, the integration of data streams into the EMS are handled separately for SCADA and WAMS. Process data are stored in different locations and formats, depending on the requirement of the applications running upon the data set. This has not only grown historically, but also follows different requirements for the databases, depending on the amount of data and processing speed requirements. Summarizing the aforementioned functions of modern control centre EMS, a state-of-the-art control centre is based on software modules which include at least the following functionalities and applications (see also [69]):

- SCADA system and database,
- Graphical user interface,
- Historical data records (archive),
- Steady state estimation,
- Information model management,
- Contingency analysis and static security assessment,
- Power flow and optimal power flow,
- Short-circuit current calculation,
- Operational planning tools, e.g., load forecast & reserve monitoring,
- Communication modules (e.g., telemetry, telecontrol and 3rd party access).

Furthermore, modern EMS comprise the following advanced functionalities for enhanced dynamic observability, assessment and decision support [69]:

- Wide area monitoring, protection, and control,
- Dynamic security assessment,
- Phasor data processing,
- Protection security assessment,
- Disturbance classification,
- Remedial action schemes, special protection schemes, system integrity protection schemes,
- Dynamic state estimation.

Driven by the need to decarbonise the energy sector, the electrical power system has reached a state of complexity that makes extensive automation functions in the operation of the power system inevitable. Since WAMS and SCADA based EMS developed independently, separate user interfaces for control room operations exist in parallel. Trends towards a deeper integration between both systems exist [71], but the lack of integration still challenges operators, and raises the probability of errors in decision making [17, 70]. Furthermore, the gap between PMU-based WAMS and SCADA based EMS functionalities has been recently described as an overlooked "*time bomb*", which prevents the adequate monitoring, protection and control of the electric power system [48].

While the quantity and the rate of gathered sensor information in the control room is increasing, the available time to react on critical situations in power systems operation decreases. In addition the number of actions to be performed during the operational planning phase increases and intraday actions are moving more and more into the control centre, closer to real time operation [101]. These changes in power system operation and upcoming requirements to include the planning time frame into the control room environment, reveals the key challenge: to set up an appropriate, consistent, and maintainable data model. The modular oriented approach helps to enable the transition towards an integrated data model, which fulfils actual and future business needs. It must support the orchestration of applications for different use cases, the integration of novel applications and methodologies and the support of legacy system components and applications. Furthermore, the future EMS needs to be designed in such way, that reaching the so-called human cognitive barrier is avoided [17, 25]. Therefore, the cognitive capabilities of human control room operators need to be assisted by appropriate tools (see also subsection 2.5).

Due to their criticality and complexity, EMS technology, always lagged behind the possibilities of IT in general, and still inherit a large potential to move away from tightly coupled monolithic systems towards open and modern modular systems [128]. Since major updates to their EMS environments involve high effort and costs, system operators conduct them only if necessary or if exceptional requirements must be met. The next generation EMS should therefore consider as many as possible future requirements, scenarios, and anticipate business strategies. To operate the increasingly complex and steadily evolving transmission system, future-proof EMS solutions must be modular to allow adaptations to changing requirements. A major obstacle to deploy advanced EMS applications and assistant functions, is in fact, that there is a lack of a reliable data basis, and validated power system models are missing. Thus, the next generation EMS must be able to provide a maintainable, exchangeable, and accurate, i.e., trusted power system model basis as a basis for advanced assistant functionalities. This gain in confidence and trust is a central requirement to enable a higher degree of power system automation.

3 Methodology

To design an EMS architecture, which anticipates processes in future power system operation, changes in regulatory frameworks and technical requirements several techniques have been applied, which are outlined in the following subsections. Starting from the explanation of the *Digital Twin* concept and its current state of deployment to real world problems, the technological requirements, modelling techniques and other methods to build a demonstration system are outlined. Finally, a future proof Digital Twin-centric EMS architecture is outlined.

Although a lot of effort has been invested to raise transparency of decisions suggested by neural networks [265, 266], a certain level of risk exists, that the decisions are biased, and may have devastating or even fatal consequences. Therefore, today's tools to secure operative security are mainly deterministic. Especially in critical infrastructures, such as power systems, which permanently change their states and topology, as well as underlying technology and control routines, stochastic methods applied by cognitive systems are challenged if the available historical training data and the real operating data differ too much from each other. One of the main reasons is that the required training data can only represent the previously observed conditions. Nevertheless, powerful tools can be developed based on methods from the field of so-called "artificial intelligence" to exploit the huge amount of data which is recorded during daily operation.

It is ultimately a question of clear observability, i.e., that all state variables are determinable, to take proper decisions in system operation. In consequence, data availability is a basic requirement for higher automation functions and must therefore be guaranteed. Therefore, the physics-based analytical modelling approach (i.e., based on *first principles*) is preferred for power system operational decision support and is therefore applied within the proposed framework.

3.1 Cyber-Physical Systems and Digital Twins

Physical processes integrating computational applications, are widely referred to as cyber-physical systems (CPS) [52]. Here, the term *cyber* refers to the involvement of computation and communication technology, which is the part of CPS allowing observability and controllability of the physical system. Thus, the electric power system constitutes a CPS in operation. It comprises field devices, substation automation systems, and control centre applications that allow interaction with the system via communication systems. The CPS perspective on power systems has been also addressed by research, resulting in several laboratory scale CPS testbeds. Demonstrator systems for research on power system control and automation systems have been described (e.g., [267], [268]). Shi et.al identified, that most existing CPS approaches have shortcomings in respecting the specific communication network and information transmission mechanisms [269]. Since most functions of modern power system control systems depend on reliable highly available ICT, the assessment of the ICT elements within the power system has grown importance and attention [270, 271].

Recently, the term CPS has been accompanied by the *Digital Twin* (DT) concept, which became a widely used synonym for virtual representations of systems, processes, or objects that are able to reflect the physical conditions by state abstraction of the mirrored instances.

The term *Digital Twin* itself can be traced back to its origin in the integrated technology roadmap of the NASA. John Vickers coined the term and introduced it within the NASA roadmap in 2010 [272, 273]. Since then, the term *Digital Twin* has been in constant transformation, but the basic concept of the DT has remained fairly stable [274]. Its appealing metaphorical strength in turn [275], and the excessive use for marketing purposes, inherit the danger that it becomes an overused buzzword [60]. This endangers the wide spread application of the promising concept, as potential users become cynical because of deceptive promises made [60]. Nevertheless, the DT has been identified by Garter as one of the top technology trends [276]. Several terms have been used synonymously, such as digital mirror, digital model or *digital shadow* [61].

The practical implementation of the concept has its origin within the manufacturing management, connecting the virtual representation of a product by information flows throughout its lifecycle [62, 274]. Following this approach, a possible application of the DT concept for autonomous manufacturing and production systems has been described in [277, 278]. An architecture for a DT-based cyber-physical production system has been described in [279]. It enables manufacturing systems to be quickly adaptable and reconfigurable to fast changing demands and to run autonomously. The opportunity to variate or reconfigure product characteristics or the production volume, without manual efforts during ongoing production is acknowledged as an economic value [280, 281]. Thus, the DT concept also allows to create a benchmark environment, although the system of interest is not yet available or existing in a certain configuration.

As illustrated in Fig. 3.1 (adapted from [61], and previously published in [282]), the main difference between a digital model and a DT is the ability to exchange data bidirectional with the physical object automatically. In contrast, a *digital shadow* only receives information by interrogation, but does not automatically exchange information. The counterpart of a digital shadow is a *digital generator* [283].

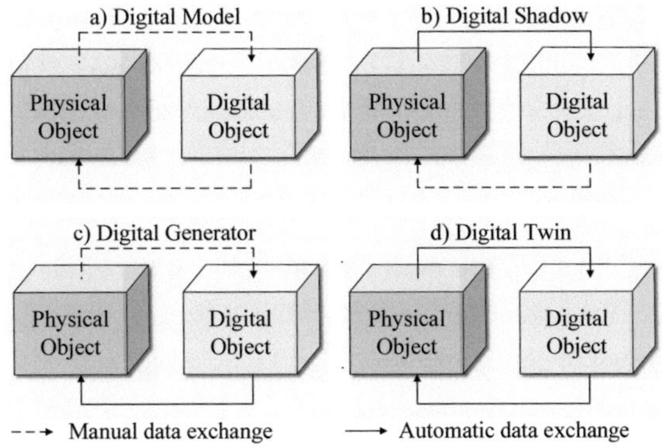

Fig. 3.1: Difference between digital model (a), digital shadow (b), and digital generator (c), and Digital Twin (d) (adapted from [61], and previously published in [282])

It can be applied for design studies, model-driven product development, or for compliance tests of control schemes prior to commissioning. This technique has been applied e.g., in the project "*Kriegers Flak combined grid solution*" [284]. The process of virtual commissioning, has been even addressed by standardisation [285]. Hence, the beforehand described predecessors of DTs are all already in practical use today.

3.1.1 The Digital Twin Concept

Since the DT concept is constantly evolving, and a myriad of publications describes the abstract DT concept and its potential applications, it is useful to provide a general description of the concept that is valid for the framework developed within this work.

As outlined in [60, 286], a model for a DT should be physics-based, sufficiently accurate and ideally real-time capable. Thus, DT is a virtual representation based on an analytical model that resolves the time-discrete states of the modelled physical system based on first principles. In contrast, data-driven machine learning approaches can only mimic the behaviour of the observed system by using a set of training data (inputs and outputs) to build surrogate models [286]. However, to improve the applicability of a DT when the analytic computational model is too time-consuming to provide insights within a required decision interval, machine learning based surrogate models are a perfect companion to a DT [282]. Based on analytical models a DT can enrich the training data, especially for unusual operational scenarios or when measurement data is not available [287]. These model characteristics allow to update model parameters p based on measurement data, and to run decision support applications within the required timescale of the process. As illustrated in Fig. 3.2, the underlying model formulation and structure can be assumed as accurate, when the measurable output y^{m} matches the response $\hat{y}^{\mathrm{DT}}$ of the model considering the same independent input variables u^{m}. This basic presumption is the foundation of the DT concept and leads to the field of DT applications, where DTs allow to interact with objects in the *physical space*, either to gather new information about the observed real-world entity or to apply advanced control functions by adaptable strategies, which have been evaluated in the so-called *virtual space* (see Fig. 3.2).

As illustrated by Fig. 3.2 a DT can derive a state value in the virtual space, which is not directly observable by measurements, and report this to the real world (physical space) e.g., for enhanced monitoring or control purpose. This fundamental property of a DT appears to be a similarity to the concept of the observer, which is often applied in control engineering. According to the control theory, observers apply a mathematical model of the observed system and a function of the measurement noise (i.e., a correction term) to estimate the system state from measurements [288]. By closing the loop, the model error can be controlled by an error function, which can help to enhance the system model accuracy and thereby the controllability of the system. Thus, if the system model is correct, and the observer's initial state equals the system state, the estimation error is close to zero along the entire trajectory. Following this explanation, an ideal DT represents the case of a high-fidelity model without uncertainties. Hence, in practical application the estimated state converges to the true state of the observed entity.

It can be concluded that the digital twin represents a special form of an observer, which provides data that can be used purposefully e.g., for future power system operation. An aspect which distinguishes the proposed DT based concept from the traditional observer approach is that it provides a higher flexibility in terms of the model basis. While the observer does only reproduce specific states of a system for control purposes, a DT can also reproduce the behaviour of the system in a predictive simulation considering changing system topology and conditions.

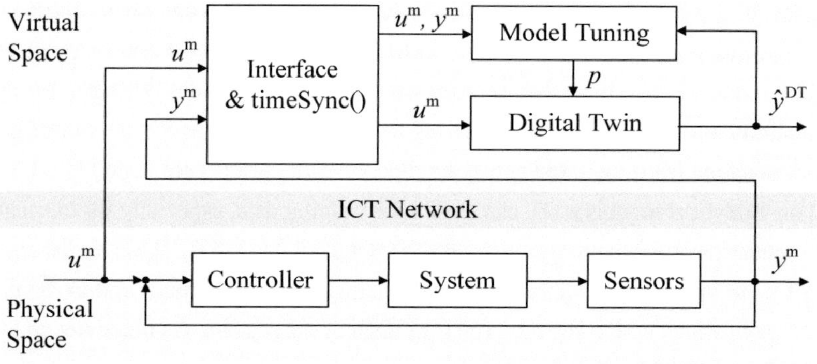

Fig. 3.2: Visualisation of a generic Digital Twin (previously published in [282])

3.1.2 Characterisation of Digital Twins

Industrial applications of the DT described in the literature, mainly describe the dimensions of the physical world, modelling in the virtual space, and necessary data exchange and services [54]. The state of the art of DT applications has been reviewed extensively in [54–56, 63, 289]. Most applications today focus on product design, product life-cycle management, autonomous manufacturing systems, prognostics, and health management. Tao et al. [54] identified the urgent need for generic modelling methods and processes, since the studies analysed do not cover equally all the dimensions involved in DT creation. Boschert et al. define the DT as a semantically linked collection of digital artefacts, which include data aspects of the design, engineering and operational data, which allows to evolve the DT along the whole life cycle of the real system [290]. This seamless integration requires an approach, which provides a framework for both consistent and flexible modelling. This approach is addressed by the German Industry 4.0 initiative by a so called asset administration shell [291], in which all information and functionalities are described in composite models (see also RAMI 4.0 [67, 292]).

In addition to the modelling aspect, an important property of DTs is their ability to contain data over the whole life cycle of the entity. This property, is referred to as *digital thread* [293]. Therefore, DTs are an integral part of the digital transformation [293], and inherit a large business value. Basic principles and definitions for life-cycle-management of systems and components for industrial process measurement, control and automation are provided in the standard IEC 62890 [294].

Frameworks and architectures for multiple purposes including the Digital Twin have been described in [279, 283, 295, 296]. DTs promise to avoid costs by recognising anomalies and device failures earlier, to fill gaps in sensor data, and to predict future operating conditions [23], [297]. Inherent potential for savings is also expected in the power system domain. These comprise operational and maintenance costs (e.g., for power plants [298]), compliance studies for implementation of new asset or controllers and improvement of quality of supply, as well as other services [299]. As their field of application is manifold, DTs possess a certain disruptive character and impact [53]. This is evidenced by market analysts who expect the business value of DTs to increase up to $48.2 billion by 2026 [300].

Due to the multitude of possible applications, a uniform definition of the DT concept is impossible, and what exactly constitutes a DT is difficult to grasp. Therefore, the main inherent characteristics of a DT derived from recent literature are pointed out:

- A DT of the object, process or system contains models of different granularity,
- A DT requires a predefinition of structure, content and purpose [301],
- A DT anticipates automatic and bidirectional data exchange [61, 62],

- A DT enables convergence between the physical and virtual states by synchronisation and appropriate information exchange [64],
- A DT can include data of all phases of the lifecycle [62],
- A DT allows to execute simulations [301],
- A DT can predict future states and supports decision making [69],
- A DT has the purpose to monitor, control and optimise [53, 302],
- A DT can develop capacities for autonomy [53].

As indicated by the discussion of the characteristics of DTs, a uniform description of the DT concept does not exist in the literature. To provide a consistent confinement for the proposed framework, the following definition is given. It comprises the relevant characteristics outlined beforehand:

A Digital Twin is a continuously adaptive digital model of a physical system, that allows automatic and bidirectional data exchange with the physical system in the real world. It has the purpose to monitor, control or optimise the system it mirrors. By learning from available data including simulation and sensor feedback, it can support decision making under volatile conditions, supports process automation and is able develop capacities for autonomy.

This general definition considering the characteristics of DTs sets the base requirements of the proposed DT-centric EMS architecture discussed within this thesis. In this context, a *Digital Twin* comprises a set of mathematical models, derived from physical laws (i.e., *first principles*) that can reflect the behaviour of the observed and modelled assets within a power system accurately. Furthermore, an interface to the observed process is a prerequisite to update states and to tune the model parameters applying sensor data.

Additional information aspects can be added as related information throughout the lifecycle of the asset represented by the DT. These may contain equipment type data sheets, reports and studies, processes or workflows. Analytical functions are applied to derive information to support decision processes. The main differences between a DT and conventional modelling and simulation methods are summarised in Tab. 3.1.

Tab. 3.1: Comparison of DT inherent characteristics with conventional simulation

Feature	Conventional simulation	Digital Twin
Communication	None, or unidirectional communication from or to the simulation model (see digital shadow and digital generator in subsection 3.1)	Allows automatic bidirectional closed-loop communication
Calculation speed	No real-time requirement	Real-time capability is intended
Data	Contains data to model specific scenarios	Contains data of the entire lifecycle
Modelling approach	Either data driven surrogate model or knowledge-based model derived from physical laws	A combination of first principle/ physical law and data driven models possible

3.1.3 Power System Digital Twins

Power system models have been used since decades for planning purposes and the investigation of system limits and behaviour to perturbations and power system planning and operation processes. As depicted in Fig. 3.3, dynamic power system phenomena have a wide range of time frames (derived from [5, 303–305]). They vary from microseconds for electromagnetic transients, such as lightning wave propagation up to slow electromechanical dynamics. Thus, the modelling techniques and requirements depend on the effects to be investigated. Furthermore, the system components itself, e.g., equipment controller configurations and their time constants must be considered. Taking power system models as an example, a multi-application high-fidelity model comprises several model representations: a steady state representation (power flow model), a phasor or average value model representation for electromechanical interactions and a model including all necessary components and controller sub-models to reflect the electromagnetic timeframes up to the microsecond scale (see Fig. 3.3).

As real-time applications in information technology and operational technology advance, the DT concept has been discovered also by the power system industry as an enabler to improve future power systems. A review about possible applications of DTs in future power systems, considering several tasks and respective model depths has been outlined by Palensky et al. in [306].

To describe the power system in all its aspects within a DT, the complexity and diversity of models requires a consistent representation considering all model classes and model depths [69, 307]. To avoid redundancy of the model causalities, a suitable meta-model (e.g., an asset administration shell) can be applied to describe the interrelated dependencies.

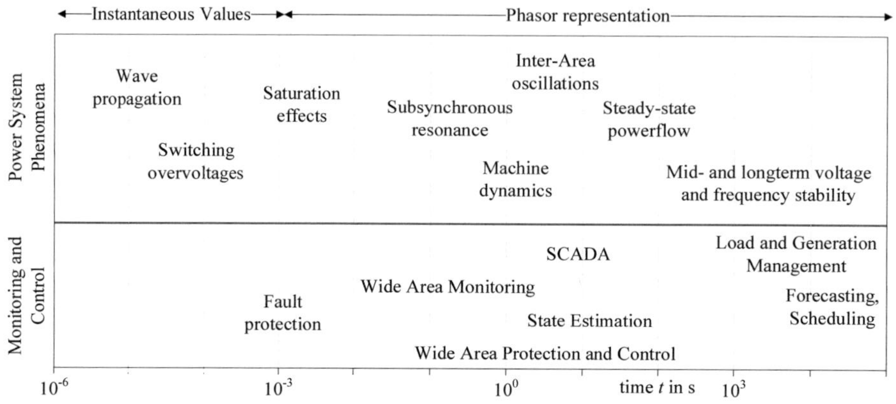

Fig. 3.3: Power system phenomena and their times horizons in terms of grid control response requirements

A holistic data storage for system operators and other power utilities, the *"Electrical Digital Twin"* has been introduced by Siemens. The software solution aims to support interoperable data exchange and synchronization of different data sources. Its purpose is to maintain grid models from different utility domains according to the single source of truth (SSOT) concept [308]. Thus, it represents an information management system, which connects and consolidates the information from different data sources to create consistent grid models. This interconnection of the different data sets and data sources aims to bundle and validate the available information and subsequently keep it consistent for all dependent applications. These also comprise power system studies (e.g., planning, operation, and maintenance). The SSOT concept allows the maintenance and synchronisation of the system model data automatically, which saves manual effort. Without the feedback from a simulation model, however, the SSOT represents only a digital shadow corresponding to the distinction given in Fig. 3.1 (see subsection 3.1). Fig. 3.4 illustrates the SSOT framework using the example of an enterprise DT [308]. The layered "automation pyramid" on the left side of Fig. 3.4 is based on the standard series IEC 62264 [309], and has been described before e.g., in [310] by Siepmann. It describes the integration of enterprise-control systems and has been developed to achieve a reduction of complexity by subdividing the processes involved in automation into individual levels. In Fig. 3.4, a SSOT framework is illustrated that represents a DT database in context of the layered "automation pyramid". It connects different processes involved in data collection and processing within the enterprise and allows to apply modelling engines and analytics to access and consolidate the data within the DT database. The metadata layer is required to maintain the data structure and to allow versioning and other database-related processes.

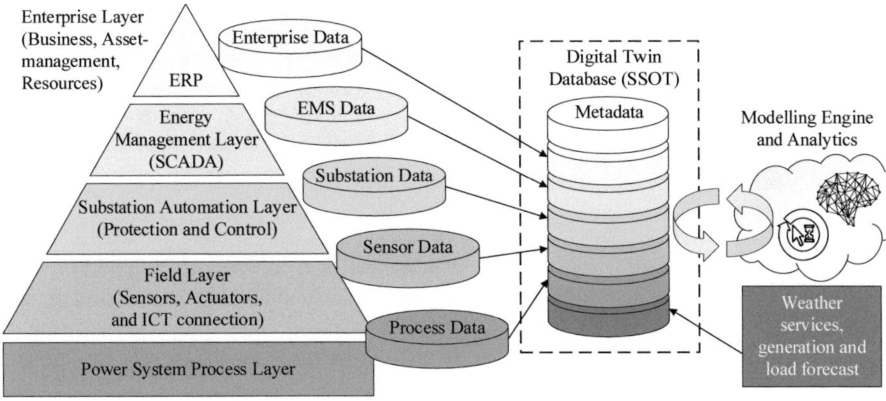

Fig. 3.4: Illustration of an enterprise SSOT applying the power system as example

An US patent [311] issued to General Electric, describes a generic application of a DT. It applies the twinned physical system to provide virtual sensor data, to predict system conditions and the remaining life time of a physical system [311]. An application of this concept has been described by GE in [298]. Here, available sensor data and analytic methods are combined to predict and to optimise power plant performance. ABB recognised the advantages of DTs for product design and optimization, as it can support investment decisions and can save time and cost, avoid waste and potentially dangerous situations when compared to physical testing procedures [312]. Applications of the DT concept for power systems are also gaining interest in academic discussions [306]. For example, a real-time hardware-in-the-loop power system simulation referred to as a DT was recently described in [313]. The scope of the research was on real-time simulation applying dynamic phasors and the advantage of the modelling approach. Its interface to other entities has the purpose to allow co-simulation, but not to connect the simulation engine to a real-world entity. This property of a DT to enable automatic and bidirectional data exchange for monitoring and control purposes is the central feature of the proposed DT-centric EMS.

3.2 Digital Twins Embedded to Energy Management Systems

The tasks of an EMS are to collect and process data to provide a fully comprehensive system state of the observed area of responsibility, which can be applied for decision making (see subsection 2.3). By mapping available process data stream to the DT model, it can serve as an additional data source and a sandbox instance for enhanced model-based automation functions. From the perspective of prospective EMS requirements, the DT concept provides the basis to develop the necessary functions to handle the power system of the future, since the required EMS functions go beyond what can be handled with today's methods, both in terms of data provision and the response speed achieved by human beings.

Approaches to utilise DTs embedded into EMS for power system operation are to the authors knowledge not yet existing in academic publications or industrial applications (besides own work see appendix A.6).

As described before in subsection 1.2, the EMS architecture as present today is not feasible anymore to enable the development of maintainable and future proof supervisory and control systems, which are flexible enough to adapt to the fast-changing requirements of system operation. Future control centre EMS solutions need to be adaptable to customer needs and applicable to operate the increasingly complex and steadily evolving transmission system. Thus, the DT concept is a perfect companion and can serve as the core instance of the proposed EMS architecture.

The core component of the novel EMS architecture is a modelling engine considering the DT concept. Therefore, the proposed EMS architecture is called DT-centric. The DT-centric EMS collects and processes data, to provide a fully comprehensive system state by mapping of real-time data stream to the DT model base. In summary, it can be considered as a special form of an observer, which provides data that can be used purposefully for future power system operation (see also subsection 3.1.2). To create a sustainable environment, it provides open interface for modular extensions, e.g., customised EMS functions or the previously described assistance system functions (see subsection 2.5), which can be supplied by different vendors.

Considering power system dynamics, the concept enables a wider adoption of synchrophasor data in control rooms, since the new type of information provided by PMU based WAMS becomes applicable by many subsequent EMS functions for decision making during real-time operation. These available measurement data streams allow to estimate the internal states of the observed system, which often cannot be obtained by direct measurement. To reproduce the system response reliably by state estimation, an accurate dynamic system model is required. Hence, in order to increase the accuracy of the estimated states and the simulation results, PMU data streams and a suitable parameter estimation method are required to enable the tuning of the power system model parameters.

Through validation by measurement data, human operators gain trust in the validity of the model and thus in the implementation of proposed decisions. Thus, instead of taking conservative decisions (considering worst case reserves), actions can be evaluated and secured by simulations. This can help to reduce stress especially in unexpected situations and mitigates the risk, that human control room operators take unfortunate decisions. To address the challenges of future power system operation, the next generation EMS must provide the following functionalities and solutions:

- Contain an appropriate, consistent, maintainable, exchangeable and machine interpretable data model,
- Provide flexibility to adapt to altering business needs by a modular software architecture,
- Support of legacy system components and applications, as well as the integration of novel applications and methodologies,
- Measures for (re-)establishing a secure network state, e.g., by setpoint adaptation of flexible asset for preventive or curative actions,
- Relieving human operators of numerous manual interventions required due to the increasing number of events during system operation and increasing response time requirements.

These requirements and functionalities have been considered in the proposed EMS architecture illustrated in Fig. 3.5. The key components of the DT-centric EMS architecture comprise the following novel components:

- A DT database including a model management platform,
- A modelling engine, that can run several model instances in parallel as a sandbox instance for decision support,
- A dynamic digital mirror which continuously validates the master model by measurands,
- A model parameter estimation function which can tune model parameters according to the objective to minimise the model error,
- An automation system, that can create scenarios from the validated master model to assess system security and predicts future system states,
- The framework enables the development of enhanced decision support applications and assistant functions to support the cognitive capabilities of human operators.

The virtual sensors and actuators in Fig. 3.5 (see also [23, 24]) can be interpreted as states that are unobtainable by measurements or not available but can be determined by the dynamic digital mirror. The proposed new components of the framework illustrated in Fig. 3.5 are discussed in depth in section 4.

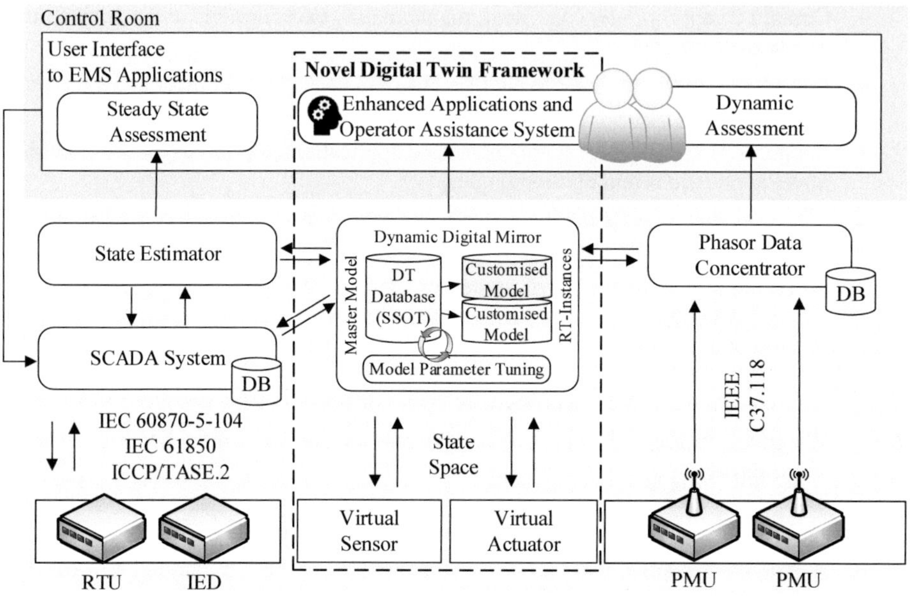

Fig. 3.5: Illustration of the proposed next generation EMS concept. The novel core components are framed in dashed lines and represent the DT-framework (see also [23, 24])

3.3 Basic Functions of Digital Twins for Operational Support

To formulate the basic functions of the proposed DT-centric EMS framework, time sequence diagrams are applied. These are based on the unified modelling language (UML) and illustrate the interactions of the objects involved during a scenario. Here, the time sequence of interactions of the objects in the physical and virtual space during a scenario is illustrated from top to bottom. Furthermore, the required functionalities of the general use cases considered within the proposed EMS architecture are illustrated. The concept to illustrate the interaction of a DT and a physical system has been adopted from [283]. To keep the sequence diagrams concise, data streams are sent directly from the process to the higher-level monitoring tools. These are illustrated in Fig. 3.7 to Fig. 3.9. They focus on the basic interacting components of a DT-based monitoring and control system, which are illustrated in Fig. 3.6.

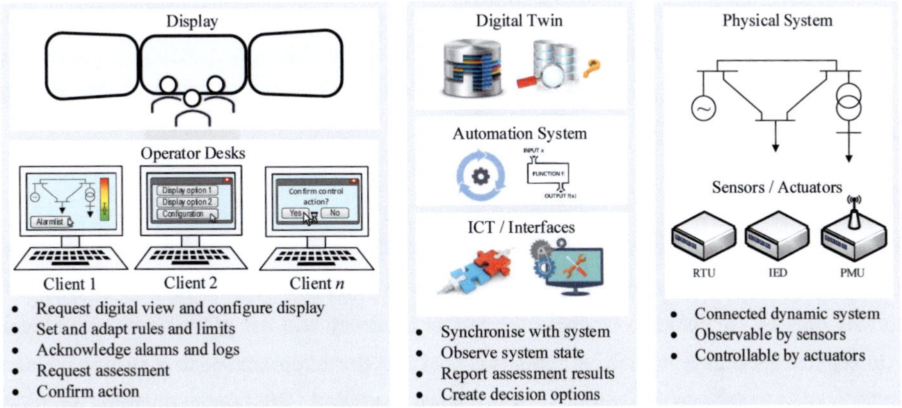

Fig. 3.6: Components of a DT-based monitoring and control system

3.3.1 Monitoring and Manual Control

For monitoring and manual control, a sequence diagram is illustrated in Fig. 3.7. In this use case the DT applies available sensor data to synchronise the inherent model to the physical system. The human operator monitors the system, interprets the information gathered by the system and manually reacts on events. If required, he can investigate effects of control actions on a DT instance before he initiates measures within the real system by service requests. In case the client requests specific assessments, the DT can create additional data. Thereby it can also serve as a proxy to obtain unobservable data to improve manual control decisions and commands. A sequence diagram, including the interacting components of a server-client DT-based monitoring system is illustrated in Fig. 3.7. As illustrated, a human operator or another external entity monitors the system, while the DT mirrors the state of the physical object continuously in discrete time intervals.

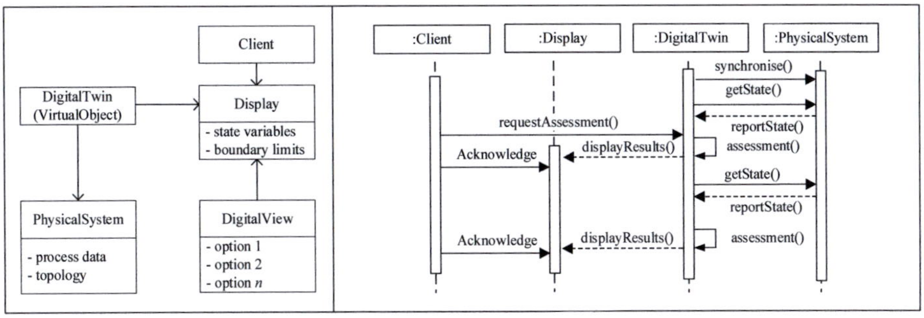

Fig. 3.7: UML and sequence diagram of a DT-supported monitoring system (see also [287])

The digital view on the display can be customised to the required context-based visualisations to improve understanding of the system state. For testing or training purposes, the client can also communicate only with the DT instead of a component in the physical system.

3.3.2 Partially Automated Control

For partially automated control, the DT adapts to the system state, which is monitored continuously by available measurements. Additional synchronisation checks at given time intervals may be executed to evaluate the model accuracy and the given confidence level. Although manual intervention by humans is possible, the automated control system handles routine tasks to relieve the human operator from workload. The human operator monitors the automation system, considers control options, and acknowledges decision options and commands of the decision support system. In the case that control intervention is needed, the control system assesses options applying the DT as a proxy to test and evaluate control commands and proposes suitable or optimal control options to the human operator. In emergency cases, when the predefined conditions are met, the automation system implements actions automatically or after acknowledgement depending on the predefined rule set. The rule-based conditional decision logic is specified and created by human operators. These interactions are illustrated in Fig. 3.8.

3.3.3 Temporary Autonomous Control

Assuming appropriate observability by measurements and DT model confidence, the system state can be predicted for certain changes and events. Based on a ruleset considering constraints of the physical system and a reliable decision logic, autonomous operation and control becomes possible for normal system conditions.

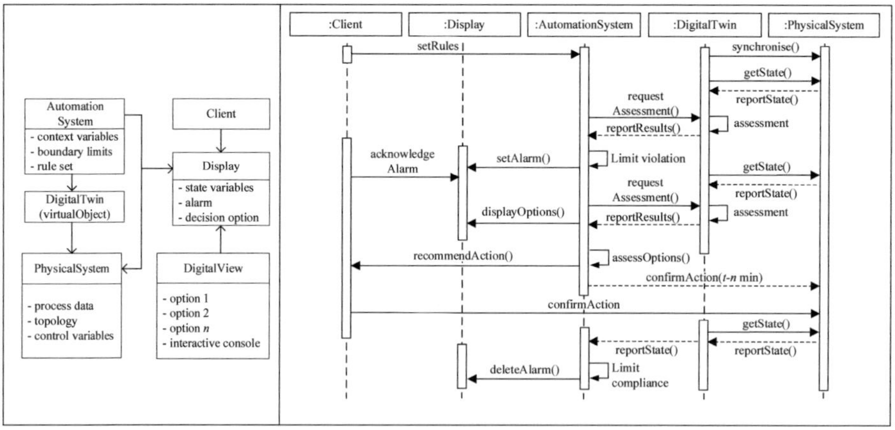

Fig. 3.8: UML and sequence diagram of a DT-supported decision support system

It is envisioned, that the physical system adapts to the previously assessed and determined optimal DT state automatically, relieving the human operator from routine control tasks as illustrated in Fig. 3.9. In case the human induced ruleset or boundary limits are violated, switching back to manual intervention mode or partial automated control should be possible to maintain system security. This includes situations where the DT model does not allow predictions, confidence in the predicted states is not given, or it does not comply with a certain level of accuracy.

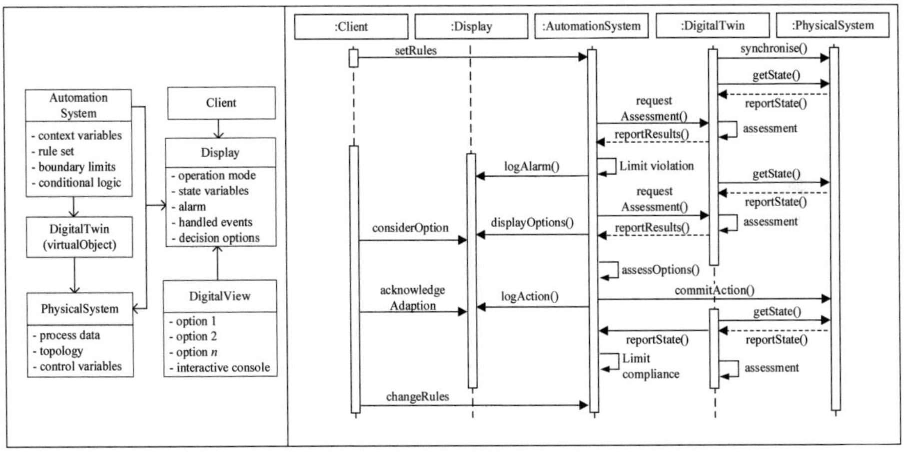

Fig. 3.9: UML and sequence diagram of a DT-supported automation system

In case of ICT failures during automatic operation mode, fallback instances, such as alternative communication systems, redundant automation systems or the manual intervention by skilled operators might be necessary. Thus, as for any automation system, the major challenge is to train human operators in such a way, that they keep engaged during automated system control periods [35, 37]. They must maintain their ability to steer the power system manually to a certain extent, and to configure and operate the automation system [35]. Since full automation, i.e. autonomous system operation must include all system conditions, such as alert and emergency (see Fig. 2.2 in subsection 2.2.2), the decision logic must either comprise all possible cases, or consist of an adaptable intelligent system. To build a future proof DT-based automation system interacting with cognitive systems is envisioned for future implementations. A cognitive system can autonomously determine its own state based on available information [287]. It can learn independently how to achieve given goals through the ability to adapt. To achieve this, methods from the field of "artificial intelligence" can be applied, but at the cost of accurate training data, since the major inherent weakness of artificial intelligence driven cognitive systems is their characteristic to adopt to biased data [287]. This must be considered when the training data and the real operating data differ too much from each other [287]. Fortunately, accurate and validated

DTs can help to enrich the training data, especially for uncommon operational scenarios. However, the power system is constantly evolving. Therefore, autonomous control is challenging to realise, and is not considered in depth as a use case within this thesis.

3.4 Model Parameter Identification

By applying dynamic power system models in the EMS, electro-mechanical interactions and stability phenomena can be investigated online. The proposed DT-centric EMS allows to keep data models consistent, update and validated by harnessing data streams. As this "living process model" included in the DT concept is not yet available, the model must be tuned during operation. An essential component of the model tuning procedure within the proposed DT-centric EMS framework is the parameter estimation function. Relevant aspects for the selection of a suitable parameter estimation method applicable for time-domain power system models are discussed in the following subsection. Based on this discussion, a moving horizon estimation method is proposed for parameter estimation in the DT framework.

3.4.1 Model Types

The basis of every simulation is a model that can approximate phenomena from physical reality. Thus, the fidelity of the models must be chosen in such a way that the simulation results allow to make a qualitative statement about the investigated problem. The challenge is not to strive for an approximation close to reality, since the modelling effort and the simulation time then increase too much. This trade-off including accuracy and precision, involves the choice of adequate model complexity and model fidelity. While simplified models may not reveal the full value of the approach, a very precise approach may lead to unbearable complexity or violate time constraints due to long simulation execution times [69].

Dynamic systems are modelled by means of discrete-time input-output models. Depending on the different properties of the system, three types of models can be distinguished: black-box, grey-box and white-box models [314]. White-box models represent a class of models, which can be described by mathematical equations whose parameters can be determined by measurements, and the model parameters typically correspond to physical parameters. Therefore, the model structure must be known in depth. Consequently, their results have a high accuracy [315]. The main drawback of white-box models is, that the effort to model complex systems is considered as high.

In contrast, black-box models avoid the extensive description by equations. Black-box models are either compiled models to protect intellectual property or, they reproduce only the input and output behaviour by applying an empirically obtained transfer function, which is derived from a measurement data of a dynamic system response. These often do not consider the internal structure of the modelled system and their parameters may not allow direct physical interpretation. These surrogate models only approximate the input-output relation of the modelled system. Furthermore, they are lacking flexibility and have disadvantages in extrapolation [314].

Grey-box models comprise characteristics of both white-box and black-box models. They are often based on an assumption about the physical structure of the process or system, but several parameters are unknown and must therefore be determined. While parametric grey-box models are built upon assumptions about the process structure and the existence of a parametric description of physical properties, non-parametric black-box models are based upon empirical weight functions and statistical distributions. As summarised by [316], the shade of grey expresses how much in depth information exists about the modelled system. The three types of models can be generally distinguished by their characteristics. These are illustrated in Fig. 3.10.

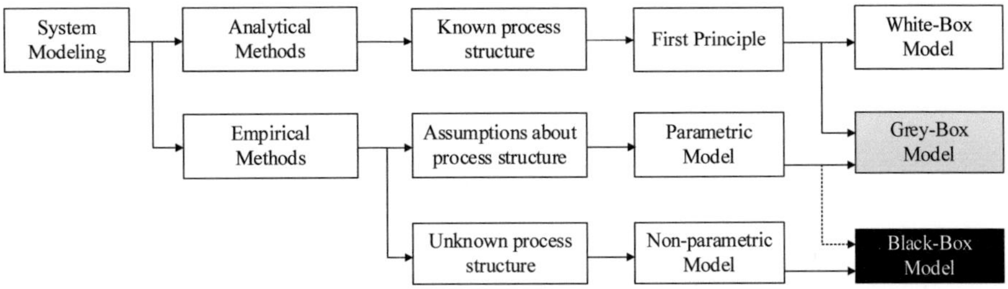

Fig. 3.10: Characterisation of system modelling approaches (adapted from [315])

Against the background of the model types used, two general model identification methods can be distinguished. While the non-parametric identification is fitting a generic model to time-domain or frequency spectrum data, the parametric identification utilises optimum search techniques to force the parameters of the underlying model to match those of the real system, comparing the output signals of the real system and the model to minimize the difference between the outputs [58], [317]. Since the model structure can be assumed to be known for EMS applications, the system parameter identification procedures of interest are based on algorithms, which apply an error function to minimise the deviation between measured and simulated output and result variables. Based on the preceding decision to apply parametric grey-box models, procedures are analysed that can be used for the task of parameter estimation of grey box models.

Model-based optimisation and control approaches depend on highly accurate models. For accurate power systems modelling, both the model structure and the model parameters must be known. For the extra-high voltage levels of the transmission system, and the high and medium voltage levels of distribution systems, the detailed topology of the system is sufficiently known. Thus, only the system model parameters must be determined.

Some EMS vendors supply static power flow models and sometimes dynamic models in case that a DSA is within the tendered product [251, 318] (see also subsection 2.3). These models are mostly derived from software solutions developed for performing offline power system studies. Therefore, online model validation is crucial before online studies can be conducted. To keep the model updated, an online parameter estimation method is required. A robust methodology, capable to estimate parameters of linear and nonlinear models is therefore proposed in the following subsection. The beforehand described modelling techniques, their characteristics and requirements for application are compared and summarised in Tab. 3.2.

The model classification given in Tab. 3.2 can also be applied to the power system industry. While power system component manufacturers often develop detailed 3-phase EMT models using the white-box model class for product development, they protect their knowledge by compiling them and converting them into proprietary grey-box or even black-box models where no access to any parameters is possible. However, these models are not openly available and require expert knowledge to implement, which results in the fact, that system operators mostly conduct power system simulations that are based on generic component models. These must be validated against measured responses of the power system components to provide valid insights into the power system behaviour. Thus, the class of grey-box models has been applied within this work, since knowledge about the system structure exists, and parametric mathematical models are available.

Tab. 3.2: Analysis of model types and their requirements for application [314], [319]

Model Type	Description	Characteristics and Requirements	Application Example
White-Box	The modelled system is accurately describable by first principles, i.e., relationships are expressible in mathematical equations.	Parameters are known and physically interpretable. Expert knowledge and experience required. Difficult to develop for large systems.	Theoretical models or detailed models of control systems.
Grey-Box	The physical structure of the system is known, but parameters are unknown.	Parameters must be estimated by observation, i.e., by measurements. The modelled system structure is known and expressible by equation systems.	Partly accessible component models. Core components may be protected, proprietary, compiled, or generic.
Black-Box	A parametric model, which adapts responses derived by measurements. The structure of the modelled system is unknown, and parameters are not accessible for any physical interpretation.	The model structure and parameters are derived through experimental modelling and are based entirely on measured data (no direct relationship to first principles). Expert knowledge and experience are not required. Disadvantages in scalability and extrapolation, and lacking flexibility.	Models, that are only required to reflect the input - output transfer characteristics of the modelled system, e.g., trained neural networks, or compiled models to protect vendor knowledge, intellectual property rights.

3.4.2 Parameter Estimation

Any system or parameter identification procedure applying observed input and simulated output data requires three main components: a set of applicable measurement data, a suitable model, and an appropriate estimation method [320]. This general statement about model parameter estimation, matches the statement about DTs given in [60]. The authors summarise, that the successful deployment of digital twins requires trust in the model, trust in the data, and trust in the algorithms applied to update the model based on the data. The statement points towards the general remark given in subsection 3.4.1 on the modelling approach based on parameterisable analytic models. Parametric models depend on internal parameters, which describe the physical behaviour of the modelled object. These parameters are often unknown and need to be estimated. This is typically done by fitting the mathematical models to measured time series data. As introduced in subsection 2.6, a few promising parameter estimation techniques exist, that allow to derive model parameters from online measurements. To estimate unknown parameters of DAE based power system models, numeric optimization is usually applied to fit the model response to observed data [316, 321].

The full vector of states from a trajectory must be considered to obtain the optimal state estimate [164]. However, taking the full information vector as input has the best theoretical properties in terms of optimality, but inherits a high computational effort up to intractability [164]. This consideration leads to the moving horizon estimation (MHE) methodology, which is illustrated in Fig. 3.11. It considers only the last N measurements from the full set of information and summarises the information outside the considered horizon in the arrival costs (see first term in equation (3.2)). Another feature is, that the MHE technique can handle constraints, which is of importance to estimate states of nonlinear systems. Thus the MHE approach combines the properties of full information estimation while maintaining a tractable online computation [164]. The general form to describe a MHE is given in [164], whereas methods for efficient implementation are elaborated in [322].

Although it is computationally more demanding in comparison to other applicable techniques, such as the Kalman Filter (KF) and its derivatives, the MHE approach is capable to deal with highly non-linear estimation problems in the presence of state constraints. By setting the horizon length N to one, the MHE becomes a one-step recursion estimator similar to the extended KF and the unscented KF [164]. Although higher estimation speed is to be expected by considering a single measurement at each sample only, the accuracy suffers accordingly and leads often to estimator divergence [164].

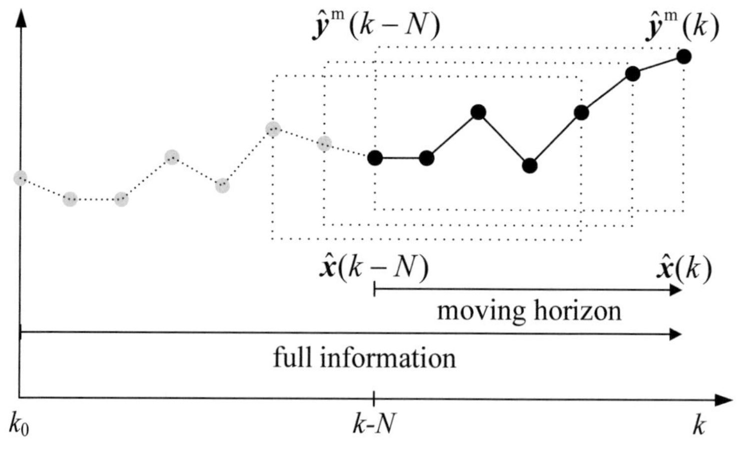

Fig. 3.11: Illustration of MHE scheme (adapted from [164])

The fundamental concept of the MHE is not to fit the parameters to the complete data available at time k, i.e., 0, 1, …, k, but rather only to the previous n time steps, i.e., k - N; $k-(N+1)$, …, k. This concept, i.e. to consider only the last N measurements and the last N values of the state trajectory is illustrated in Fig. 3.11 [164].

As the time frame is moving along the captured time series, the objective function is minimised to estimate the parameters which fit the model response to the captured values. At each time step k an optimisation problem is solved to minimise the objective function. Starting from an initial parameter set, the model parameters are varied in such a way that the objective function (3.1) reaches a minimum. To obtain a minimum, the method of least squares for the objective function is applied due to its convex characteristic. Thus, the determined minima are global minima if the selected set of functions to be solved is convex. The MHE objective function J is defined by (3.1) and (3.2). The first term of (3.2) represents the arrival cost, the second penalises the deviation of the computed output $\boldsymbol{y}$ from the discretely measured output $\hat{\boldsymbol{y}}^{\mathrm{m}}$, and the third penalises deviations from the system dynamics and enforces the parameters to behave like in steady state.

$$\min_{(\hat{\boldsymbol{x}},\hat{\boldsymbol{p}})\in X^{N+1}\times P^{N+1}} \quad J(\hat{\boldsymbol{x}},\hat{\boldsymbol{p}},\hat{\boldsymbol{u}},\hat{\boldsymbol{y}}^{\mathrm{m}})\in \mathbb{R}^{\mathrm{d}(N+1)}\times\mathbb{R}^{\mathrm{p}(N+1)} \tag{3.1}$$

$$\begin{aligned} J(\hat{\boldsymbol{x}},\hat{\boldsymbol{p}},\hat{\boldsymbol{u}},\hat{\boldsymbol{y}}^{\mathrm{m}}) &= \left\| \begin{matrix}\hat{\boldsymbol{x}}(k-N)\\ \hat{\boldsymbol{p}}(k-N)\end{matrix} - \boldsymbol{\chi}^{k} \right\|_{\boldsymbol{S}}^{2} \\ &+ \sum_{n=k-N}^{k} \left\| \hat{\boldsymbol{h}}(\hat{\boldsymbol{x}}(n), z(\hat{\boldsymbol{x}}(n), \hat{\boldsymbol{u}}(n)) - \hat{\boldsymbol{y}}^{\mathrm{m}}(n) \right\|_{\boldsymbol{R}}^{2} \\ &+ \sum_{n=k-N}^{k-1} \left\| \begin{pmatrix} \hat{\boldsymbol{x}}(n+1)-\hat{\boldsymbol{x}}(n)-\Delta t\boldsymbol{f}\left(\hat{\boldsymbol{x}}(n), z(\hat{\boldsymbol{x}}(n)), \hat{\boldsymbol{u}}(n), \hat{\boldsymbol{p}}(n)\right) \\ \hat{\boldsymbol{p}}(n+1)-\hat{\boldsymbol{p}}(n) \end{pmatrix} \right\|_{\boldsymbol{Q}}^{2} \end{aligned} \tag{3.2}$$

The power system model (see also subsection 4.3.1) can be described by differential algebraic equations (DAE), i.e.,

$$\dot{x}(t) = \boldsymbol{f}(\boldsymbol{x}(t), z(t), \boldsymbol{u}(t), \boldsymbol{p}) \tag{3.3}$$

$$0 = \boldsymbol{g}(\boldsymbol{x}(t), z(t), \boldsymbol{u}(t), \boldsymbol{p}) \tag{3.4}$$

$$y(t) = \boldsymbol{h}(\boldsymbol{x}(t), z(t), \boldsymbol{u}(t), \boldsymbol{p}) \tag{3.5}$$

where $\boldsymbol{x} : [0, \infty) \rightarrow \mathbb{R}^{\mathrm{d}}$ represents the differential states, $\boldsymbol{z} : [0, \infty) \rightarrow \mathbb{R}^{\mathrm{a}}$ represents the algebraic states, $\boldsymbol{u} : [0, \infty) \rightarrow \mathbb{R}^{\mathrm{c}}$ describes the control inputs, $\boldsymbol{y} : [0, \infty) \rightarrow \mathbb{R}^{\mathrm{o}}$ describes the measurable outputs, and $\boldsymbol{p} : [0, \infty) \rightarrow \mathbb{R}^{\mathrm{p}}$ collects all parameters.

To estimate the model parameters, the state vector $\boldsymbol{x} = (\boldsymbol{x}^T, \boldsymbol{p}^T)$ is augmented, i.e., the parameters $\boldsymbol{p}$ are formally treated as states without dynamics (i.e., $\dot{\boldsymbol{p}}(t) = 0$), as done in [323].

Furthermore, the DAE system described by (3.3), (3.4), and (3.5) is discretised with step size $\Delta t > 0$ such that equations (3.6) to (3.8) are fulfilled for all $n \in \mathbb{N}$.

$$\hat{\boldsymbol{x}}(n+1) = \hat{\boldsymbol{x}}(n) + \Delta t \hat{\boldsymbol{f}}(\hat{\boldsymbol{x}}(n), \boldsymbol{z}(\hat{\boldsymbol{x}}(n)), \hat{\boldsymbol{u}}(n), \hat{\boldsymbol{p}}(n)) \tag{3.6}$$

$$\hat{\boldsymbol{p}}(n+1) = \hat{\boldsymbol{p}}(n) \tag{3.7}$$

$$\hat{\boldsymbol{y}}(n) = \hat{\boldsymbol{h}}(\hat{\boldsymbol{x}}(n), z(\hat{\boldsymbol{x}}(n)), \hat{\boldsymbol{u}}(n), \hat{\boldsymbol{p}}(n)) \tag{3.8}$$

To incorporate physical constraints of the differential states x and the algebraic states z, lower and upper bounds can be included i.e., $\overline{\boldsymbol{x}} \leq \boldsymbol{x} \leq \underline{\boldsymbol{x}}$, and $\overline{z} \leq z(x) \leq \underline{z}$ where $\boldsymbol{x} \in \mathbb{R}^{\mathrm{d}}$. Solving (3.2) repeatedly in a moving horizon fashion leads a new estimate for the parameters $\boldsymbol{p} = \boldsymbol{p}(k)$ at each time step $k \in \mathbb{N}$. The MHE algorithm is illustrated as a flowchart diagram in Fig. 3.12. As illustrated a DAE model is required as an input to the algorithm. Furthermore, the discrete step size $h = \Delta t > 0$, the time instant $k \geq N$, the past N+1 control vectors $\boldsymbol{u}$, the weighting matrices $\boldsymbol{Q}$, $\boldsymbol{R}$, $\boldsymbol{S}$, and the discrete time measurement data $\boldsymbol{y}^{\mathrm{m}}$ ($N \in \mathbb{N}_{\geq 2}$) are required. The vector $\boldsymbol{\chi}^{\mathrm{k}} \in \mathbb{R}^{\mathrm{d}(N+1)+\mathrm{p}(\mathrm{N}+1)}$ denotes the guess of initial differential states and the parameters based on the previous time step. The weighting matrices $\boldsymbol{Q}$, $\boldsymbol{R}$, and $\boldsymbol{S}$ are hyper parameters that can be applied to tune the convergence of the estimation procedure. Note that the MHE algorithm illustrated in in Fig. 3.12 continues the estimation procedure even if there is no feasible parameter found for a certain time horizon. Since updating of the model parameters is usually not necessary all the time, the previously found valid parameter set can be retained until a significant deviation between model response and time series data is detected.

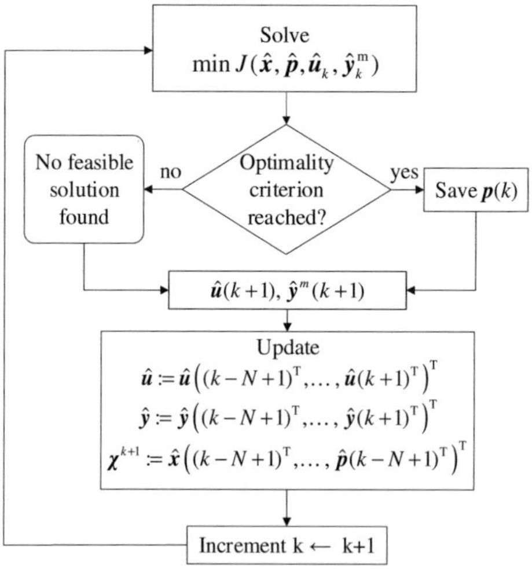

Fig. 3.12: Flowchart diagram illustrating the MHE algorithm

3.5 Conclusion on the Methodology

The methodologies described in section 3 comprise an analysis of the DT concept from the perspective of research and industrial applications. It can be stated from the existing applications, that a DT is always an individual solution pointing to the respective enterprise and process requirements.

Due to the large area of possible application, and the steadily growing literature within this active field of research and development, a uniform definition of the DT does not exist. Therefore, a consistent definition has been provided which serves the purpose to develop a future-proof DT-centric EMS. By responding dynamically to changes in the environment, DTs provide a major opportunity: to make better informed decisions during system operation.

The basic functions considered within the proposed DT-centric EMS have been illustrated by sequence diagrams in subsection 3.3. The interaction of the objects in physical and virtual space are shown. Hereby, the scope was set on the interaction between the user, the DT and the automation instance. The cases monitoring and manual control up to a certain level of automatic or even autonomous control were discussed. These basic use cases set the outline for the proposed DT-centric EMS architecture described in subsection 3.2.

The quality of the power system process model has a significant influence on the decision-making process. Especially the trust into the generated decision and response options is strongly related to the accuracy of the system model. Today's approaches to tune a power system model are dominated by manual interventions. Automated model tuning and synchronization between system model and the controlled physical device is not given.

The keystone of the DT concept is a validated high-fidelity model. To provide an online updated model, an appropriate procedure for model parameter tuning is required. A functional method, namely the MHE has been proposed for this purpose. The method is chosen by the requirement to fit time series sensor data to a parametric nonlinear model based on DAE as usually applied in the power system domain. To avoid dismissing the DT concept as hype or a marketing buzzword, this section aims to propagate understanding of the DT concept by providing an overview of its application in power system operation and planning, as well as the transformative nature of the concept for power system companies and operators.

The DT concept epitomises the convergence of mature information and operational technologies. Due to its manifold advantages, a successful deployment of the concept in the power system domain is only a matter of time.

4 Proposed Modular and DT-centric EMS Architecture

In the preceding sections, the state-of-the-art in technology and research, the need for a modular maintainable EMS and the requirements of a future-proof EMS were described. Based on these findings, a novel framework is proposed, that takes advantage of the DT-concept. After a short description of the proposed DT-centric framework for the next generation of EMS, a requirements analysis is carried out. The key components of the framework and a test setup to validate its architecture are presented hereafter.

4.1 The Digital Twin Framework

A software framework provides the foundation for independent application development. Thus, a properly designed framework allows to create an integrated but modular software system. Each component within the framework is furthermore accessible via open interfaces. This implies that applications are running separately and can be developed independently from the framework itself. The modular design approach creates the basis for a vendor independent development process, while maintaining interoperability [287]. Another paradigm within the framework is to keep data and applications separate. This entails the advantage that the applications can be updated without the risk of data loss. The principal architecture of the proposed DT-centric EMS framework is illustrated in Fig. 4.1.

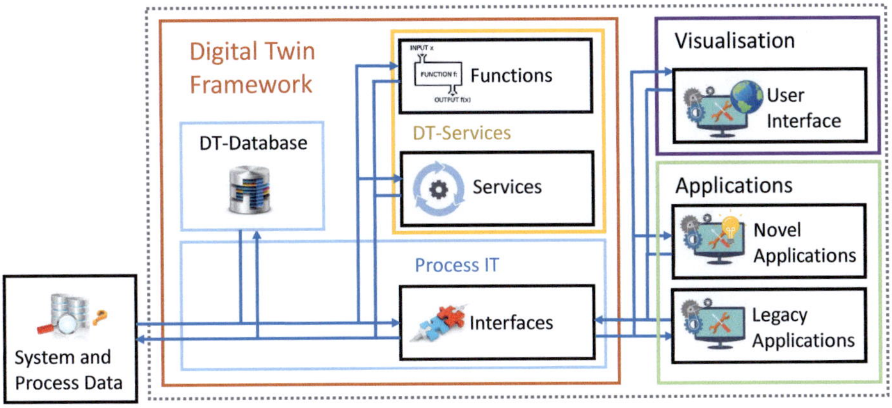

Fig. 4.1: Simplified architecture of the proposed Digital Twin framework (see also [287])

It comprises the following components:

- A DT database, containing structured information and data (e.g., models and archived sensor data),

- A process IT, connecting all internal and external components (e.g., sensors and actuators, applications, and the visualisation for user interaction),
- Required functions and services to create a DT (e.g., data model validation and calibration tools, a simulation engine, and analytical functions).

As illustrated in Fig. 4.1, the process data is captured by ubiquitous sensors deployed in the observed power system. The DT-database aligns and links model and sensor data. The services and functions within the DT framework provide the necessary applications to analyse the data, and to validate the DT model and to run simulations. The process IT provides the connection between all components. A non-exclusive selection of basic functions is summarised in Tab. 4.1. The requirements to create a power system DT to support EMS applications will be discussed in the following subsection 4.2.

Tab. 4.1: Selected components and their functions within the DT framework

Component	Purpose or task	Example Function or Application
Process IT	Bidirectional data transfer	Measurement data transfer (telemetry)
		DT model synchronisation
		Command transmission (telecontrol)
	Interfaces/ transaction manager	Data and information object provision
		Interface to applications
		Management of application interaction
DT-database (SSOT)	Central and consistent data source	Model maintenance and management
		Data consistency check
		Component model library
		Data archive
DT services and functions	Model creation	Model instantiation
	Model validation	Model accuracy assessment
	Scenario builder	Creation of scenarios on demand
	Simulation engine	Look-ahead simulation
		DSA sweeps
		Contingency screening
		Optimization
	Model tuning	Model parameter estimation
	State estimation	Steady state and dynamic observability
		Model calibration
	Topology processor	System topology tracking
	SCADA system	Processing of SCADA telegrams
	PDC and PDP	Processing of PMU data streams

4.2 Requirement Analysis

As described in the preceding subsection 4.1, the DT framework consists of several components. These have specific tasks assigned to them and are analysed in the subsections 4.2.1 to 4.2.6. The specific requirements to implement the proposed framework into a DT-centric EMS concept are investigated as well in the following subsections.

4.2.1 Information and Communication Technology

For monitoring and control applications, ICT systems are indispensable. It manages all internal communication and provides the connections between the process and components which require data inside the DT framework. It provides specific interfaces, which enable the applications to access data stored in the DT database or from the observed process.

As real-time processes in power system operation have a direct impact on system security, they are of critical importance. Their availability is the key requirement for reliable system operation. Thus, ICT systems are bound to strong security requirements and high reliability demands. However, transmission and distribution grid operators, have fundamentally different tasks, which also impose different requirements on the communication network. Depending on these requirements and economic constraints, other communication technologies, such as the unidirectional ripple control technology or mobile telecommunication systems (e.g., LTE, UMTS or GPRS) are utilised in addition to their own fibre optic networks. These communication links have a lower availability and bandwidth; thus, they are not recommended to operate system-critical processes. Future vertical communication using IEC 61850 from substation to control centres may rely on the guideline IEC TR 61850-90-2, to enable a seamless integration into the IEC 61850 standard series [324].

Modern communication networks of TSOs and DSOs are often based on fibre optic cables and are usually owned by themselves to exclusively transmit data within the enterprise. The IT security concept of the system is usually based on measures to maintain perimeter security i.e., access limitations and user roles in addition to firewalls and demilitarized zones. The IT infrastructure required for the operational processes is designed with redundancy. This applies to both the control centre process IT system, as well as for communication channels connecting the substation equipment and field devices (see subsection 2.2.5).

As the ICT required for the proposed concept is available and can be considered as secured, the data quality and data availability are of particular importance. Both aspects are discussed in the following subsections.

4.2.2 Data Quality

Data quality can be divided into the aspects accuracy, latency and completeness [325]. The accuracy of data expresses processing errors and the time skew introduced by transducer or the data acquisition methodology. The latency summarises all delays between sensor output at the interface of a measuring device and availability to the application or user [133]. The data completeness is a measure to describe data package losses.

Accuracy

The accurate reconstruction of incidents using simulation models depends on good data quality. Thus, adequate sensors for monitoring and event recording are required. Power system monitoring sensors have in common, that voltage and current transformers are applied. Due to nonlinearities in the transducer, the accuracy strongly depends on the characteristic of the transducer's burden. Additionally, the measurement accuracy depends on the sampling algorithm at the analogue to digital converter interface. The accuracy of modern sensors applied in power systems varies between ±0.01 % and ±2 %, depending on the technology and field of application [326]. The signal to noise ratio (SNR) for PMU data streams has been investigated in [327]. It has been shown that the SNR can be approximated by a value of 45 dB. To improve the measured signal quality, filters can be applied in a preprocessor to mitigate the most disturbing effects.

Latency

The causes for communication delays, latency, or jitter can have different origins. They differ depending on the communication technology and the transmission distance. Communication latency can affect controllers, thus a decisive negative influence on the stability is to be expected, especially on closed control loop control systems [328]. In general, connection-oriented communication, e.g., TCP protocols inherit an increased latency in comparison to handshake free communication (e.g., UDP protocol), or sampled values broadcasts as realised by the IEC 61850 GOOSE. The encryption of the data leads to an increase in latency.

Considering the latency of PMU data streams, the IEEE standard C37.118.2-2011 [209] gives a rough time range. The basic communication delay is approximated with 3.4 µs/km to 6.0 µs/km, while further data processing (e.g., error detection or correction routines) can add up to several hundred milliseconds [209]. A typical PMU to PDC communication is expected to have an achievable delay, within the range between 50–130 ms, including measurement data processing and computation delays [19, 209]. Experimental investigations have shown, that the cumulated latency for centralised applications can be expected within a range of 150 to 220 ms for monitoring, and is below 500 ms for control applications

[329]. Special protection schemes considering transient stability, including load and generation tripping, have the shortest latency time requirement among EMS transmission grid applications of approximately 100 ms [330].

The recommended timings for communication and information exchange between remote terminal units and intelligent electronic devices inside and outside a substation are documented in the standards IEEE 1379 [331], and IEEE 1646 [332]. A selection of relevant timing requirements for applications within the substation, and external to the substation are documented in Tab. 4.2. Further performance requirements for SCADA measurement based services are given in [133].

Tab. 4.2: Maximum delivery time requirements for selected types of information [332]

Information Types	Substation internal	Substation to EMS
Protection Information	5-7 ms (1/4 cycle)	8-12 ms
Monitoring and Control Information	16 ms	1 s
Operations and Maintenance Information	1 s	10 s

Bandwidth

As stated in [332], the IT network utilisation resulting from real-time traffic in substations should not exceed 15 percent of the theoretical maximum capacity of the network for 2 ms to avoid ICT network congestions. This requirement was also considered in the IEC 61850 standard series (see also 2.2.5), to achieve guaranteed real-time capability for protection-critical applications. According to a survey on bandwidth and latency demands for transmission grid applications [330], the average bandwidth required for modern transmission system applications is within the range of 5–10 Mbps for telemetry within the observed control area and between 25–75 Mbps for communication between control centres.

Data Loss

In contrast to electricity transmission, in information technology it is possible to discard the transported information unit (i.e., to drop a package) in the event of excessive data transfer requirements or priority conflicts. This property of ICT must be considered when designing closed loop control systems, where information must reach the intended instance within a certain amount of time.

4.2.3 Database

Traditional EMS comprise a SCADA database (see subsection 2.3.1). It stores process data in the form of data points. Thus, it represents a data master, without detailed structural information describing the supervised system. To derive a higher degree of observability, a state estimator is required (see subsection 2.3.2). It applies a topological model to estimate the system state, resulting in a steady-state representation of the power system. To further increase the degree of observability, all available process data sources are required. In consequence, the database requirements are increasing. On the one hand, it must comprise the basic functionalities of a SCADA database to maintain a consolidated steady-state data set. On the other hand, it also must be able to process higher resolving real-time data streams. This integrated solution also referred to the SSOT concept as described before in subsection 3.1. It provides an information management system that connects different data sources. Thereby, it can maintain and consolidate all available models, comprising offline and online power system studies. Furthermore, it should provide secured interfaces to other departments in the enterprise (see subsection 3.1.3).

4.2.4 Model Management

As outlined in subsection 4.1, an extendible, structured, and maintainable data model is required, to create and maintain a Digital Twin. The CIM/CGMES standard (see subsection 2.2.5) inherits these requirements and is thereby dedicated for structured data modelling and model exchange in the power system domain [333–335]. It furthermore allows to track incremental changes. This allows to map different scenarios or model versions consistently in a common database. The CIM/CGMES standard provides an extendible modelling framework containing different model classes, including their attributes, and topological relations. A drawback is, that the publicly available CGMES profiles contain only generic models, which require special attention, when getting parameterised, and need to be validated during online operation (see also subsection 3.4). The CGMES profiles relevant for this work are explained in Tab. 4.3.

Tab. 4.3: Examples of CGMES profiles (see also [23])

Profile and abbreviation	Description of contained objects
Equipment (EQ)	Equipment and container objects of the data set
Topology (TP)	Switching state and resulting network topology
Steady State Hypothesis (SSH)	Objects required to exchange input parameters needed for a power flow calculation
State Variables (SV)	Results from last valid power flow solution
Dynamics (DY)	Information required to model power system dynamics
Diagram Layout (DL)	Instructions for creating a network graphic
Geographical Location (GL)	Position of equipment in geographical coordinates

4.2.5 Functions and Services

The DT functions and services of the DT framework have been described in Tab. 4.1. These comprise a modelling engine, the SCADA and the PDC and PDP respectively to process incoming telemetry data or outgoing telecontrol messages. Furthermore, a topology processor to track topological changes, a state estimator, and a model tuning instance are required. The latter keeps the parameter set of the DT model updated. The most of these subroutines and services are already part of modern EMS today and have been described in section 2.

The novelty compared to today's approaches is the deep integration of the data sources with the DT. This includes the automatic model tuning and the synchronisation between the DT model and the observed power system. Assuming the existence of a DT model, which is continuously updated, other task specific models can be inherited from the DT master model. These can be executed within a dedicated simulation engine to provide insights about the actual or prospective future system state. All these processes require fast or real-time data processing. Here, the term real-time refers to the processing of data by a computer in connection with another process outside the computer according to the time requirements of the external process [336]. This includes systems, operating in conversational mode, or processes, that can be influenced by human intervention while they are in progress [336].

4.2.6 Summary of the Requirements

The requirements to create a power system DT have been analysed. As the purpose of the DT is to support power system operation, it strongly depends on a reliable ICT system. Thus, ICT availability and integrity will gain importance in future. In power systems, where many communication interdependencies exist, the bandwidth requirements rise in comparison to the present needs. This is of critical importance especially in cases where a higher degree of power system automation is in scope. The objective of the envisioned DT-centric EMS is to keep a consistent and accurate master model for subsequent applications. To achieve a high model accuracy, a high quality of the incoming measurements from the distributed sensors is required.

Another important factor for successful implementation is an efficient data and information management. To achieve automated model maintenance, suitable database concepts are required to create maintainable and extendible data models.

To create a reliably performing EMS, the underlying process IT must be highly available, and capable of processing a high number of data streams. The definition of uniform interfaces allows to organise applications in a modular manner and reduces the required time and engineering effort during the implementation phase.

4.3 Design of the Test Environment

A framework to build a DT-centric EMS has been described in the previous subsections. The requirements of the framework have been discussed and its capability to provide the foundation to develop dedicated applications for future power system operation has been outlined. The novel key component of the DT-centric EMS is the dynamic mirror model, that adapts the power system model to the dynamic responses of the physical power system. The challenge here is to evaluate the deviations by an appropriate measure of quality. Furthermore, the model must be valid for different system configurations, system states and operational scenarios. Therefore, a flexible and modular modelling engine has been implemented. Furthermore, the model to be tuned must be available in an accessible format so that the parameter estimation method can be applied (see subsection 3.4).

An extensive test as commonly done in the industry for such systems is not possible within this research work. Only the feasibility of the proposed DT-centric EMS can be proven (see also subsection 4.5). Nevertheless, to demonstrate the core functionalities of the proposed framework, a test setup comprising the core functionalities has been set up. The applied test environment comprises the following components:

- A data source to create synthetic process data streams, realised by a dynamic reference simulation set up with DIgSILENT PowerFactory,
- A structured model library, comprising the required equipment models,
- A time-domain simulation engine applying suitable numerical integration routines,
- A topology processer which is enabled to track state changes (e.g., load steps, shunt or transformer tapping, topological changes, and setpoint changes),
- An interface to connect simulation data streams with the mirror model instance,
- A parameter estimation method for model tuning from the process data streams,
- Visualisation functions.

The test environment applied for the research and development in this thesis is illustrated in Fig. 4.2. It consists of a process data source, modules for data exchange (SCADA signals and PMU data streams), and functions for data processing. The topology processor tracks state and topology changes from the SCADA system to provide the correct switching state to the model instance. These transitions from one to another state are triggered by events during the runtime of the time-domain simulation. The applied modelling engine, namely the dynamic digital mirror (DDM), comprises the power system model, based on the last observed system state and assumed parameters of the underlying component models. The model design complies with the CIM/CGMES standard and is created in the manner of a composite framework. The DDM is MATLAB based (see subsection 4.3.1) and capable of solving DAE based power system models in the time-domain. The complete DT-centric EMS framework furthermore includes a transaction manager (see subsections 4.3.4 and 4.4). It coordinates the information exchange between the components illustrated in Fig. 4.2.

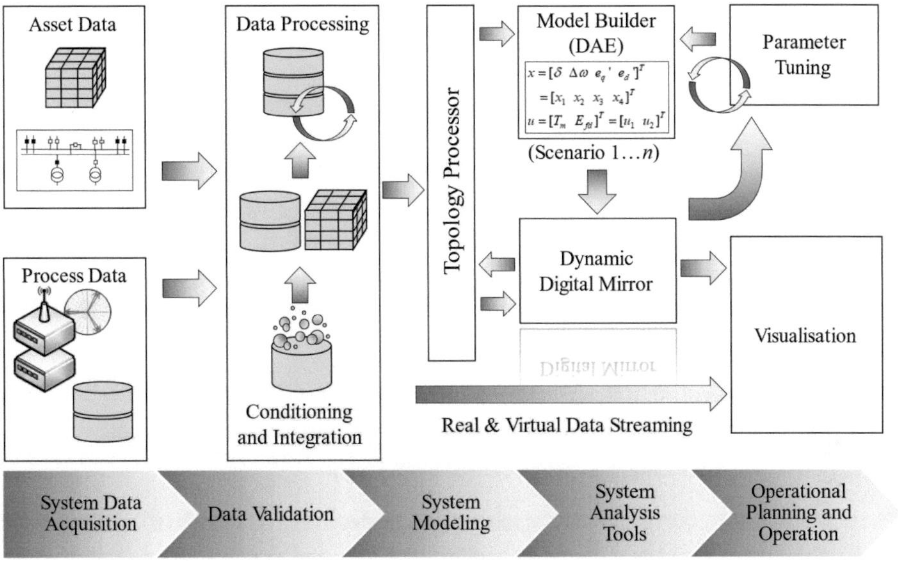

Fig. 4.2: Basic structure of the demonstration and test system (see also [24])

4.3.1 Power System Modelling Engine

As outlined before in subsection 3.1.1, a dedicated modelling engine is required to create a functional DT. To model the power system in the time-domain efficiently, a fundamental assumption is that the electrical frequency of the power system remains nearly constant [5]. This allows to express sinusoidal voltage and current quantities as fundamental frequency phasor. In transmission system modelling, furthermore the assumption that a balanced three phase network exist, reduces the computational effort significantly. These assumptions allow to represent the system quantities as single-phase positive sequence phasors.

Although a fully integrated solution of a power system DT may also comprise the modelling electromagnetic transients, these interactions are not in focus considering real-time power system operation. In contrast to electromagnetic transients, which mainly effect system dynamics locally (e.g., protection devices), the electromechanical transients affect the whole system and are observable by PMUs. For these reasons, DSA applications also deploy a phasor representation. The sample rate of the PMUs data streams corresponds well to the simulation step size applied in the phasor domain [217, 337]. The main advantage of phase domain modelling is that in a modern EMS the PMU data streams are available as input for model validation. Furthermore, the phasor representation is sufficient to investigate relevant stability criteria [5, 338] and the interactions between the HVAC and HVDC grid [96]. In the phasor domain, the synchronous machines and their controllers dominate the dynamics of the HVAC system [339, 340].

The advantages of a phasor representation of the electrical power system quantities can be summarised as follows:

- Dynamics of large power systems can be computed efficiently,
- PMU data streams can be used directly to tune the system model dynamics due to the same discretization timeframe in the range of milliseconds,
- The results are easy to interpret and visualization as well as comparison of real measurements and simulation results is straight forward,
- Electromechanical dynamics are used for dynamic stability and security assessment in today's EMS applications; thus, operators can adopt their knowledge to the new system.

The major components of the power system model are transmission lines, loads, transformers, generators, converters, controller models, busbars, and terminals. The latter represent nodes that connect the conducting equipment electrically.

The dynamic interaction of these components can be described by means of differential and algebraic equations, which are summarised in a DAE system (see also subsection 3.4.2):

$$\dot{\boldsymbol{x}} = \boldsymbol{f}(\boldsymbol{x}(t), \boldsymbol{z}(t), \boldsymbol{u}(t), \boldsymbol{p}) \tag{4.1}$$

$$0 = \boldsymbol{g}(\boldsymbol{x}(t), \boldsymbol{z}(t), \boldsymbol{u}(t), \boldsymbol{p}) \tag{4.2}$$

$$\boldsymbol{y}(t) = \boldsymbol{h}(\boldsymbol{x}(t), \boldsymbol{z}(t), \boldsymbol{u}(t), \boldsymbol{p}) \tag{4.3}$$

Whereas the vector $\boldsymbol{x} \in \mathbb{R}^n$ represents the differential states, the vector $\boldsymbol{z} \in \mathbb{R}^n$ represents the algebraic states, $\boldsymbol{u} \in \mathbb{R}^n$ describes control inputs, $\boldsymbol{y} \in \mathbb{R}^n$ represents the measurable outputs, and $\boldsymbol{p} \in \mathbb{R}^n$comprises all parameters. The differential and algebraic states and the resulting response of the system are described by the functions $\boldsymbol{f}$, $\boldsymbol{g}$ and $\boldsymbol{h}$.

To describe the interconnected power system, independent equation systems which describe the models of the elements and equipment are composed to create a model of the complete system. The equipment models, which can comprise sub-models (e.g., controllers) are then linked by defining the output variables of one as input variables of the other. The system model can include further components such as reactive power compensation systems, converters, or other types of power plant models.

The electrical grid itself, represented by the nodal admittance matrix $\underline{\boldsymbol{Y}}$, is defined by the branch properties (i.e., transmission lines, cables, transformers) and their topological interconnection. The load models are integrated into the system model accordingly. They can either be integrated by shunt elements as passive loads into the nodal admittance matrix, or they are integrated as dynamic loads via a separate DAE sub-model. In Fig. 4.3 (adapted from [341]) these composite sub-models are shown schematically as an interconnected system model.

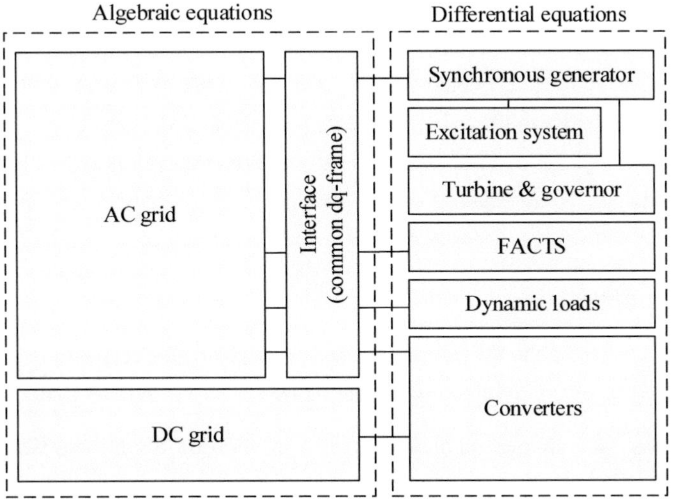

Fig. 4.3: Structure of the composite power system model (adapted from [341])

Time-domain simulation

To digitally compute a dynamic simulation in the time-domain, a time-continuous system must be represented as a time-discrete system. Thus, a direct derivation or integration of the mathematical equations of the time-continuous system is not possible. An approximated solution of the dynamic behaviour by suitable methods needs to be derived. Therefore, the differential equations to describe continuous-time systems are expressed as discrete-time systems by difference equations and solved numerically. Many numerical integration methods have been developed to solve discrete-time systems. Details about these methods can be found in the literature e.g. in Chua and Lin [342], Press et al. [343], or Butcher [344].

The explicit modified Euler is often applied for numerical integration due to its advantages of simple implementation [345]. The algorithms applied are described in detail in appendix A.2.3. A flow chart of a numerical integration routine shown in Fig. 4.4. White boxes represent steps applied within the numerical integration routine (adapted from [346]). Grey boxes represent extensions and interfaces required for an EMS integration. The process data interface illustrated in Fig. 4.4 is required to track system state changes during the simulation. It enables the application of events, i.e., to consider changes of state in the simulation within the next simulation time step. The sequence of events can be considered by the corresponding time stamp. Thus, the network model is constantly updated, and a complete data set is available at any time instant. The main stability loop can be executed in a parallelised manner, to investigate several scenarios in parallel simulations (see also Fig. 4.8 in subsection 4.3.5). The DDM modelling engine (see also subsection 4.3) can set up a specific scenario, comprising specific component models. These comply to the

CIM/CGMES standard (see subsection 2.2.5) and are maintained in a database as depicted in Fig. 4.4. Here t represents the actual time and h indicates the time step size of the numerical simulation routine. To determine whether the power system is stable or not, a time-domain simulation is conducted. The system is unstable if the simulation diverges, i.e., the trajectory of the phasors is not asymptotically stable. According to [347], the accuracy of this methodology *"only depends on the numerical integration method and on the system model"*.

Real Time Simulation

Considering applications for power system operation, which rely on real-time processing of simulation results, the following requirements can be formulated [348], [23]:

- A real-time simulation must provide a solution for a time step before it expires in real-time,
- Delays in the processing of input and output variables must be small,
- For transient simulations, the time step needs to be selected in such a way, that the underlying device controllers are represented well and react accurately.

Running dynamic simulations faster than real-time, allows operators to "look-ahead" and to take dynamic contingency analysis into account for operational decisions [349]. Depending on the complexity or size of the simulation model, the computational burden is a challenge especially for large power systems. For transmission systems, the modelling complexity can be reduced by considering the following simplifications [5], [23]:

- The electrical frequency is assumed to be nearly constant, i.e., complex voltages and currents can be represented by phasors,
- Impedances and lumped transmission line models are applied instead of elementary components, i.e., voltages and currents are assumed to change instantaneously,
- Reduce the number of differential equations in the power system model (e.g., by application of low order models or reduction of the number of dynamic devices).

Besides model simplification, several other approaches are applicable to speed up power system time-domain simulations [23]:

- Use of powerful computation hardware and apply parallelisation,
- Improve algorithms to speed up the approximation of numerical solvers,
- Increase the integration step size of the numerical routine, or apply time-step variable solvers,
- Apply factorization for efficient handling of the network admittance matrix,
- Reduce the model order, i.e., the number of differential states to be determined in the model,
- Reduce the stiffness of system equations by avoiding very large and very small time-constants within the same model instance.

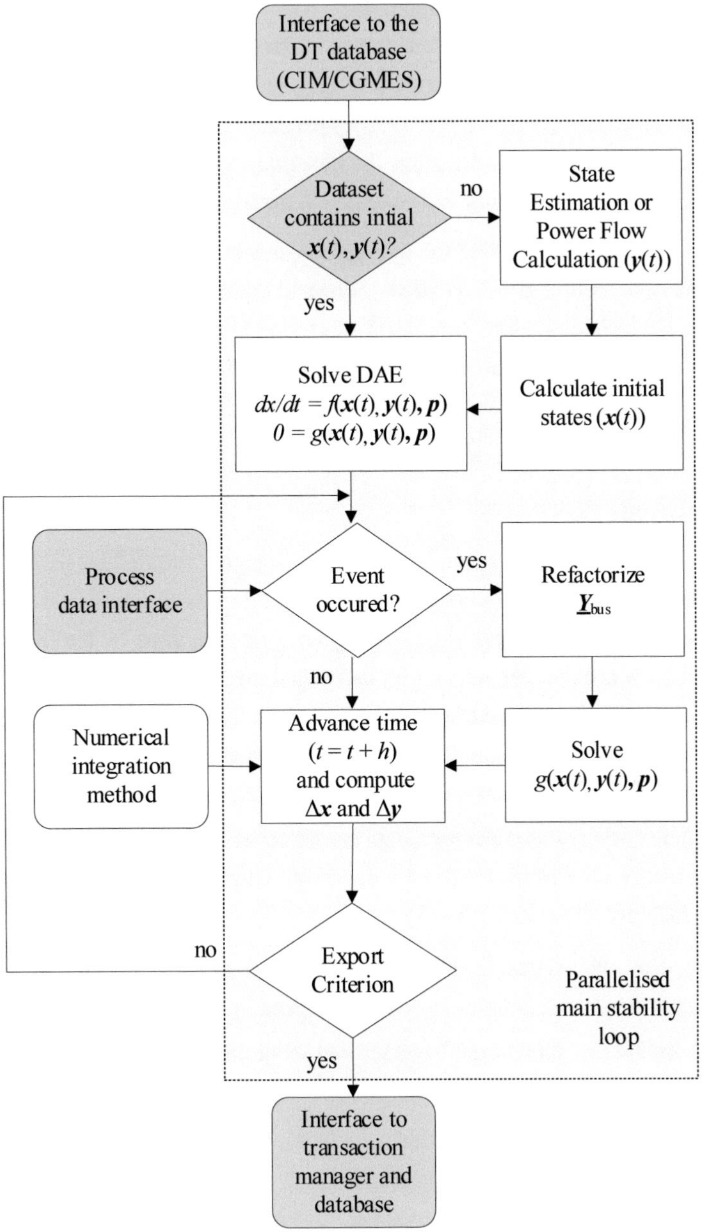

Fig. 4.4: Flow chart of the numerical integration routine for power system time-domain simulations (adapted from[346]). The grey boxes illustrate the additional components to interface the EMS environment (see also [23]).

The last key point imposes a challenging task: Since converters controls in comparison to other devices have typically small time-constants, the system models containing converters tend to contain stiff system equations, which require special attention to solve them within a reasonable time. A valid solution is to apply parallelisation, and to exchange the relevant measurement and control signals. Thus, the dynamic simulation is split into several parts, that can be solved independently and executed in parallel (see also subsection 4.3.5).

An additional assumption is, that the system operates in balanced conditions. This allows the application of a single phase (positive sequence) representation. This simplification was selected for the validation of the approach (see section 5). It should be noted, that the chosen phasor approach does not imply any undue restriction of generality for the entire approach to build a power system DT, but is merely intended to illustrate the validity the proposed concept.

4.3.2 Topology Processor

The topology processor tracks topological status changes, reported by SCADA messages to determine, and update the model representation of the power system automatically. The initial topological state is established via a general interrogation by the SCADA system. Subsequent state changes are tracked by individual messages according to IEC 60870-5-104, to keep the network admittance matrix up to date. These topological state changes and load changes (when modelled as static loads) are considered in the system models network-admittance matrix $\underline{\boldsymbol{Y}}_{\text{bus}}$. If the communication fails, state changes can no longer be tracked directly. Some missing status changes, e.g., in case of malfunctions of a single device, could be substituted by state estimation, from observing changes in the power flow or related measurements (e.g., from PMUs).

To track topological changes in the power system, the model network-admittance matrix $\underline{\boldsymbol{Y}}_{\text{bus}}$ is altered. As the dynamic models are interfaced to the network-admittance matrix to obtain algebraic states, those must be refactored according to topological changes (see also subsection 5.1.5).

The events are considered during simulation in the real-time model. They represent state changes and are processed in the following order:

1. The event is recognised by the topology processor,
2. The event is assigned to the corresponding variable in the model,
3. The event is executed within the next time step in the simulation routine.

To model certain logics within a simulation (e.g., protection device settings), a special class of events, so called state events [350] are applied. These are executed in the case that specified state variable conditions are fulfilled during the simulation.

In contrast to the consolidated bus-branch model illustrated in Fig. 4.5 a), a full-topology node-breaker model as illustrated in Fig. 4.5 b) is initialised in EMS. It allows the tracking of switch positions, and to keep the network topology updated in real-time. The node-breaker model can be converted to a simplified bus-branch model representation. However, a reverse transformation is not possible directly, due to the loss of necessary information. An additional advantage of the detailed node-breaker model is the simpler reproduction of contingencies.

Within an EMS a CIM conform model is maintained. This is considered within the design of the concept. This standardised representation of power system models enables the flexible exchange of topological data (see subsection 2.2.5) and is expected to be available within electric utilities.

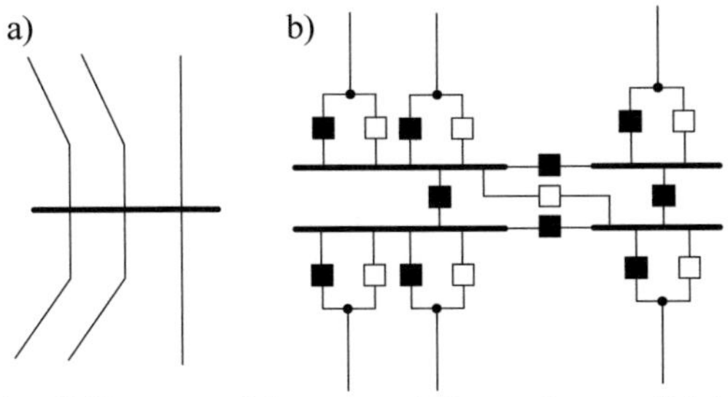

Fig. 4.5: Example for different model representations: a) consolidated bus-branch model, b) detailed node-breaker model of a busbar comprising circuit breakers, as well as lengthwise and crosswise bus-bar couplings (isolators are not considered)

4.3.3 Model Validation and Model Parameter Tuning

To obtain information about the model accuracy, a model validation function is required. In case a given model can reflect the dynamic system response well, it is considered as validated high-fidelity representation of the observed system. If deviations between correctly obtained measurands and simulated data exist, either the model itself is wrong (e.g., model structure, model type, model component or model fidelity), or its parameters need to be adjusted. In the latter case (see Fig. 4.6), the comparator function validates the model component responses by comparison to the measured signal $\hat{y}^{\mathrm{m}}$ and the model response $\hat{y}^{\mathrm{DT}}$ within a feedback loop. The model parameter set $\boldsymbol{p}$ is aligned by the identification function until the absolute error is within satisfactory limits. To derive accurate model parameters, the sensors capturing the measurements are ideally located at the terminal of the corresponding modelled equipment. The model structure is known by nature for the applied use cases within the EMS (see grey-box models in subsection 3.4.1).

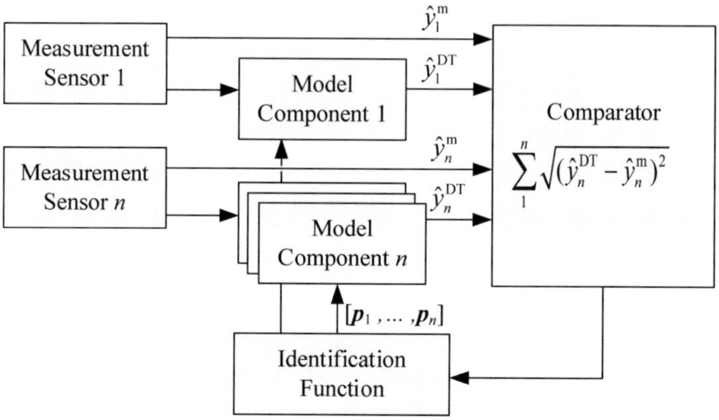

Fig. 4.6: System parameter identification principle, based on [351], (see also [58]).

Thus, the chosen method (see subsection 3.4.2) allows to estimate model parameters for component models as required for the DAE based DDM described in subsection 4.3.1.

It should be noted that any model can only approximate the measured behaviour. Thus, the model parameters which deliver the best fit between the dynamic model and the measured response, do not necessarily have a real or physical meaning.

4.3.4 The Transaction Manager

From the perspective of information flows, an automation system interconnects many different components. While only small, but time-critical amounts of data are usually transmitted at the actuator/sensor level, the amount of exchanged data increases significantly at the system control level. These systems are typically server-client based, and hard to maintain in cases where many components are interconnected. To support the integration of a rising number of data streams and the creation of flexible systems, middleware approaches have become established [352]. These event-based streaming solutions are also referred to as a transaction manager. The transaction manager coordinates, schedules, and processes the information and data exchange related operations required in the EMS. A conceptual description of the transaction manager and the underlying concept, is given in [353] and [354]. Analogous to operating systems for computers, it considers the available ICT resources (e.g., available memory) to organise the processing in an optimal way in terms of availability and data integrity. In comparison to a conventional database, the transaction manager does not necessarily need the relational descriptions of objects and their attributes. The concept bases on state changes, i.e., it is event-based and stores these state changes as a log entry in a persistent real-time stream. The transaction manager consists of a service-oriented architecture, connecting data sources, e.g., databases, sensors, and software applications by event streams. These event streams are stored for later access and can be manipulated and processed. Routing the event streams to the destination services ensures a continuous data flow. As illustrated in Fig. 4.7, services and information are connected by the streams, which are organised as topics. These topics can be accessed by a publishing or subscribing service, whereas the transaction manager (e.g., a Kafka Cluster [353]) connects these. An advantage is, that the role assignment of the active components are flexible, i.e., subscribers can also be producers via another topic, whereby the logical path is maintained. Thereby, the transaction manager extends the communication towards a higher flexibility. Depending on the configuration in practice, it allows direct communication between corresponding information publishers and subscribers, i.e., sensors with data repositories or applications, and automation functions with actuators. In contrast to today's established solution, previously calculated states and information are thereby available on demand and can be subjected to a more detailed analysis if required. Keeping information available in the transaction manager's topics is a very performant solution in comparison to conventional database interactions. Relevant data, which is no longer needed on an ad-hoc basis, can either be stored persistently, withdrawn, or assigned with an expiry date. The data required in a persistent manner, e.g., consolidated models or measurement data is stored and maintained in the DT database.

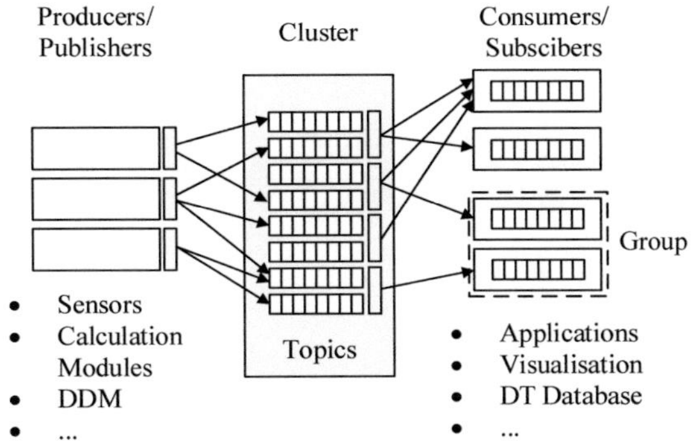

Fig. 4.7: Principle of a transaction manager as EMS streaming platform (adapted from [354])

4.3.5 Parallelisation and Data Exchange between Model Instances

As described before in subsection 4.3.1, the modelling engine supports the execution of parallel simulations. This allows to investigate several scenarios at the same time or to screen several scenarios at once. Furthermore, the co-simulation of models becomes possible. This can be of advantage when models of different detail level in large power system models are simulated or different numerical solvers are required for certain model components (e.g., to solve stiff DAE systems). The co-simulation of scenarios and data exchange between model instances during runtime is illustrated in Fig. 4.8. To exchange the data between model instances, a uniform description of the algebraic and dynamic state variables is required. If this preliminary condition is met, the routine to initialise dynamic state variables from the latest state estimation or power flow result can be omitted.

Another relevant use case is, to get a snapshot of the current simulation and run several scenarios from this state, without the need to initialise each model instance. The publisher subscriber approach also allows to alter conditional variables or system states, while the simulations are executed. This powerful instrument enables the screening for contingencies, and to adapt the search objective, in case the system model must be adapted to react on state changes. A drawback is, that model instances must wait for a specific simulation result to proceed itself.

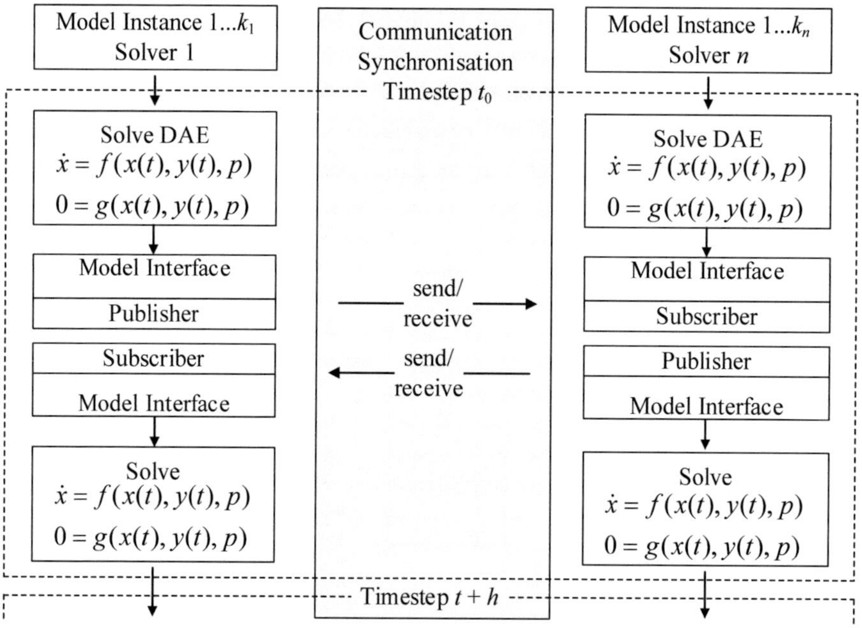

Fig. 4.8: Co-Simulation of DDM instances using a publisher and subscriber approach (partly adapted from [355])

4.4 Reference Architecture for Implementation in Practice

A future-proof EMS must support the orchestration of applications for different use cases, including the support of legacy system components and novel applications, organisational changes, and iterative growth. Therefore, it should be created in a modular manner. To provide a concept for practical implementation of the proposed next generation EMS, a reference architecture has been developed, which is outlined in this section. As the DT framework is embedded into the EMS architecture and represents the central element, the architecture is called a DT-centric EMS. It is illustrated in Fig. 4.9 (see also [23]).

The EMS architecture points towards a consistent data modelling and universal access to the available data. Therefore, the application modules of the proposed EMS architecture are separated from the core calculation functions and the DT database and model management system (see also subsections 4.2.3 and 4.2.4). The modular approach allows to develop software functions independently from a software vendor and the support of legacy EMS applications. It furthermore enables a possibility for retrofitting of advanced assistant functions, e.g., cognitive systems or functions that guide grid operators though system operation.

The concept can be extended in such way, that all related power system models within the enterprise (e.g., in the planning department) are available to the model management system and the functional application modules. This enables the creation of a consistent model throughout the live cycle of assets and the corresponding asset models.

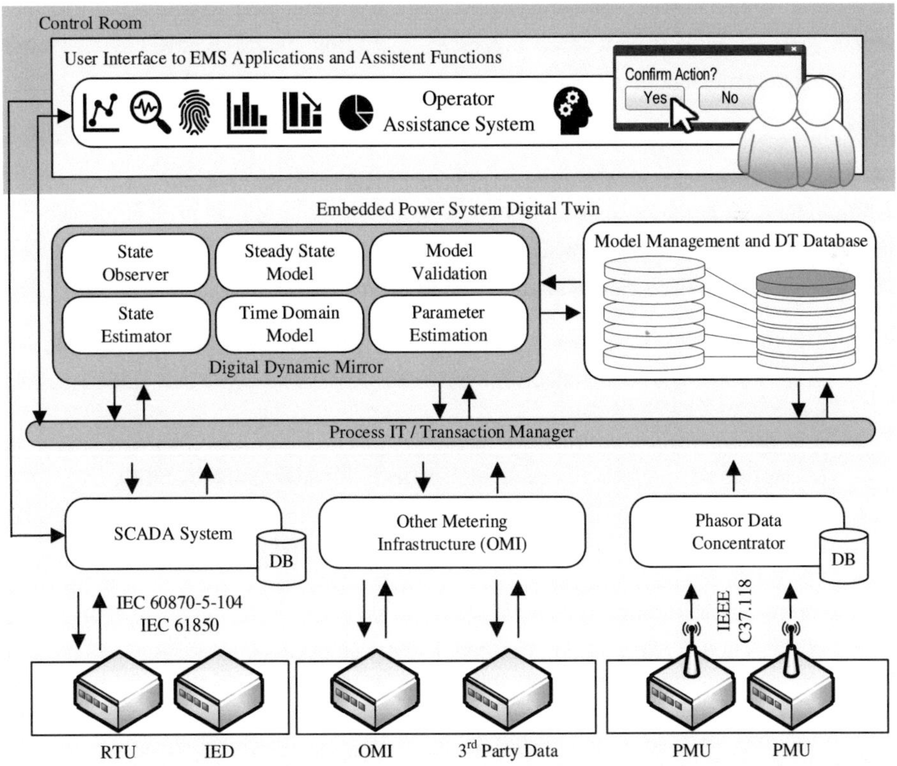

Fig. 4.9: Illustration of the proposed next generation EMS architecture (adapted from [23])

The transaction manager based on a middleware approach, allows a flexible management of the data transfer between all interacting components required to operate the power system. It handles the information objects and data streams, within the EMS and controls the process communication between data sources, modelling instances and analytic functions.

The modular EMS architecture envisions an independent operation of the telemetry and telecontrol functions. Thereby, the widely accepted standardised communication protocols described in subsection 2.2.5 can remain in service. These comprise conventional SCADA protocols and PMU data streams. As illustrated in Fig. 4.9 interfaces to other or subordinate TSO/DSO network operators, metering infrastructure or third-party data (e.g., prognosis services or weather forecasts) also exist.

As the power system evolves towards a higher complexity, automated controls will gain importance to maintain security boundaries and system stability. To generate control options and prove the decisions before implementation to the real power system, a validated model is required. The proposed EMS therefore comprises a model validation function and the dynamic digital mirror modelling engine. Thus, the model basis to obtain an observable system, is transformed from a purely algebraic description towards a power system model containing dynamic model components, which are represented by DAE.

The validated power system model together with a real-time simulation platform can be applied as a sandbox environment, which allows to generate decision options from predictive simulations. Thereby a higher degree of automation on the system level is enabled. The novel components and features which distinguish the proposed modular DT-centric EMS from a conventional EMS can be summarised as follows:

- The embedded DT represents the central element in the next generation EMS,
- A central DT-database, that allows universal access and consistent data modelling, throughout the enterprise,
- A transaction manager handles the communication and data transfer between all interacting components to enable higher flexibility and maintainability,
- A time-domain modelling approach is considered to anticipate dynamic security assessment, dynamic boundaries and power system stability,
- The DDM modelling engine serves as sandbox instance to generate and prove the control options before implementation and enables a higher degree of automation,
- SCADA and WAMS based information allow enhanced observability and are kept accessible for model validation and parameter estimation,
- The modular architecture supports vendor-independent development of software modules and enables the retrofitting of advanced assistance functions.

The DDM modelling engine is a central component of the EMS. A concept for implementation into the EMS is illustrated in Fig. 4.10. The DDM instances for "look-ahead" simulation can be executed in parallel to solve a higher number of scenarios in the C2RT timeframe (see also subsection 2.2.1). The grey boxes in Fig. 4.10 illustrate interfaces to the EMS environment, while the white boxes are DDM internal functions to handle the time-domain model.

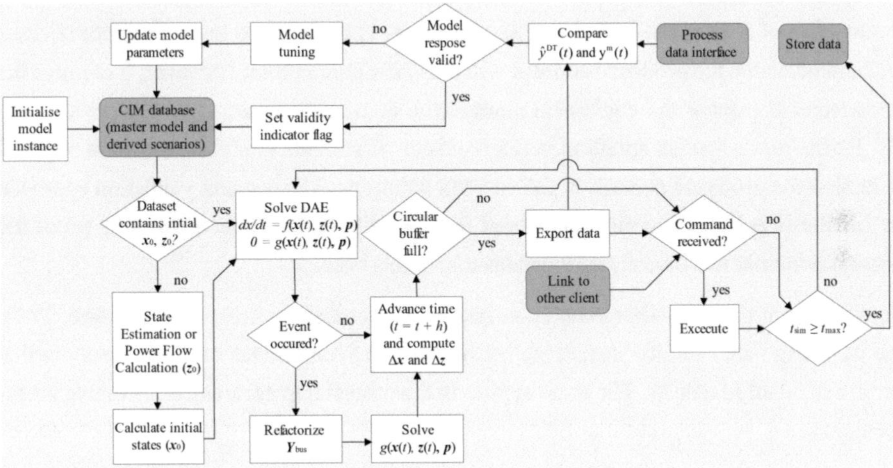

Fig. 4.10: DDM instances for "look-ahead" scenario screening and model validation during online system operation (partly adapted from [356])

The CIM database contains the master model, which is validated by process data and tuned if required. As efficient model parameter tuning relies on dynamic state transitions in observed the power system, a PMU based WAMS is a basic requirement. The online model validation can be applied following a state transition (e.g., following a power generation schedule change, or a load step) to reassess the model parametrisation and retune the model.

The modelling engine comprises a circular buffer to handle the requirement of different simulation times and step size requirements. It ensures, that the data generated is available to generate decision options during the simulation by attached analytic functions. These higher automation functions and options for manual intervention are represented by the generic link to other clients.

4.4.1 Obstacles for Implementation

The time constraints imposed by real-time operation require modelling assumptions that reduce the simulation effort to a tolerable level. These computational extensive simulations require dedicated computation platforms. Although, the scalability of such computation platforms is given to reach real-time performance, the implementation of such systems within EMS environments is not yet common practice in the energy sector. This obstacle can be overcome, but it requires expert training and capacity building.

DSA studies inherit high computational costs for large power system models. To reduce the simulation effort, a significant reduction can be achieved by filtering contingencies of interest for dynamic assessment to screen only for relevant contingencies.

Today, the number of standardised CIM compliant models is limited to generic models. As the number of standardised CIM compliant models is expected to grow in future releases and amendments, the proposed concept will also gain momentum. However, it requires the knowledge of experts and equipment manufacturers to create customised equipment models. Furthermore, special attention in the provision of suitable model interfaces is required to exploit the proposed concept in future EMS solutions. The test and validation approach in this thesis is limited to models derived from the literature. Thus, it can only proof the general principle and investigate mechanisms on this basis.

The quality of the parameter estimation results is impacted by measurement noise. Thus, the incoming data from the measuring instruments and transducers need to comply with a certain standard of quality. The same applies to the time-stamp accuracy of incoming measurements.

If suitable sensor data of high quality are available, it becomes possible to tune the model parameters according to the observed dynamic response. A drawback is that severe disturbances that include the most valuable information to tune model dynamics, are scarce and not supposed to occur during power system operation. Furthermore, to reproduce the event within a simulation, the model validation function requires exact knowledge of the event that occurred to be able to adjust the simulation model accordingly. This obstacle can be addressed to a certain extend by applying available techniques, such as the topology processor or event classification techniques, which apply the specific signature or pattern of events.

4.4.2 Limits of the Approach

The application of generic models from available literature, can only offer a limited insight. In consequence, without the application of real power system data the results can only lead to generic conclusions. To guarantee deterministic solutions and to derive viable decision options, the application of specific models is necessary. In the case that an exact model description is unavailable, the dynamic behaviour can only be approximated. Since component manufacturers incorporate intellectual property (e.g., control algorithms) in their specific models, strict non-disclosure arrangements are required to gain an applicable model description.

To generate valid decision options within the C2RT timeframe in system operation, real-time capability is a fundamental requirement. Accordingly, when schedules in system operation become tighter, the performance requirements also rise [41]. Furthermore, the computation of large power system models may results in long simulation times. In particular, the approach is only suitable to a limited extent for distribution networks, as the models of converter interfaced generation often require a smaller simulation step size and asymmetric three-phase modelling to provide the necessary information for decision support.

4.5 Approach for Validation of the Conceptual EMS Architecture

A particular difficulty is to set the scope for the validation of the functionalities of the framework presented in this thesis. The question arises, how to evaluate an architecture, especially in comparison to an existing one? From this point of view, it is a fortunate decision to base the proposed concept on standardised solutions (e.g., CIM/CGMES data model, standardised ICT solutions applied for power system automation, etc.). The validation therefore focusses only on the novel EMS components and the relevant interaction of these components to provide the promised functionality. The validation approach consists of two parts:

First, the DT itself, i.e., the virtual model is validated by proofing that the DT model matches the dynamic responses of the physical power system for several scenarios. This can be proven by comparing the outputs of the physical system, i.e., the measurements and the virtual model's simulation results. Since a demonstration of the concept with a physical power system has not been possible within this research work, a process simulation created with the well-accepted power system simulation software DIgSILENT PowerFactory has been applied. To validate the core functions of the concept, the test environment is instantiated to model dynamic system responses and to improve the model accuracy e.g., to improve the trust and confidence into online dynamic security assessment (see also subsection 2.3.3). The validity of the test environment is proven by comparing the outputs of the process models result, i.e., the measurable variables and states and the virtual model's simulation results. This model validation procedure is demonstrated in subsection 5.2.1. To update the dynamic model parameters based on available process data, the MHE method for parameter estimation from time series data is examined in subsection 5.3.

The second part of the validation approach comprises the evaluation of the novel EMS architecture. Conceptionally, at least all functions described in the state-of the art analysis of subsection 2.3 and 2.4 as well as the basic functionalities described in subsection 3.3 need to be validated by dedicated usability tests. In the industry, these comprise user feedback, and reliability or stress tests, i.e., all parts of an acceptance test as conducted for any operational system. Within this work, these extensive tests are not feasible, especially because the conceptual architecture has only been set up to a limited extend to demonstrate core functionalities. To obtain an impression of the capability and to validate these aspects, numerical simulations are carried out (see subsection 5.2). These correspond to scenarios that are also relevant in real power systems. The implemented test platform comprises the fundamental features of the conceptional DT-centric EMS described beforehand. These functionalities are validated in section 5. The tests comprise the following aspects:

- Run simulations offline and online (i.e., in parallel to the process),
- Provide enhanced observability by reflecting dynamic power system states,
- Apply process data for model parameter estimation,
- Provide information for decision support,
- Create a sandbox environment, where e.g., remedial actions can be assessed before deployment,
- Enhance trust and rise confidence into simulation results and decisions derived therefrom.

4.6 Conclusion to the Design Approach

The concept of modelling power grids with high-fidelity analytic models has existed for decades before the DT concept has been developed. As the DT concept integrates many of these existing tools and techniques, it can be regarded as next evolutionary step for real-time simulation and analysis (see also subsection 1.2). Embedding the DT concept into an EMS as core component, led to the proposed DT-centric EMS architecture presented in section 4. The requirements, the design approach as well as obstacles for implementation in practice have been discussed.

An essential prerequisite for the DT-assisted EMS is an appropriate and validated power system model. Besides the apparent application for system operation in the control room, the outlined DT-centric EMS framework facilitates automated maintenance of the power system model for various process steps from power system planning to operation, as well as the interaction with the market, so that calculations and decisions are made based on uniform models and data sets.

Today, human operators working in the control centre rely on experience to perceive and react to situations which endanger the power system security. With a DT model of the power system and appropriate algorithms, control strategies and decision support options can be developed completely outside the view of the operators and at a speed far faster than the cognitive ability of humans.

Here the purpose of the DT within the EMS becomes evident. It supports the automation of the increasingly complex power system operation, so that decision making itself is possible under challenging time constraints. A method has been introduced, to keep the power system model valid. Thereby, the accuracy of the model can be increased, which raises the confidence in the model-generated decision options and allows to operate the power system closer to its physical limits while maintaining secure operating points.

The real-time capability of the conceptional functions remain the main obstacle for online implementation. It lies in the nature of numerical simulations especially for large analytical models with a high number of state variables are bound to a high computational effort. Whether it is possible to solve the DAE for several scenarios within a limited time frame depends on the modelled details, the system size and, of course, the computing platform. Several approaches to speed up simulations have been therefore discussed. Considering an appropriate computational platform and certain permissible modelling assumptions, the real-time capability can be ensured to support online power system operation. Finally, an approach for validation of the proposed EMS architecture has been described. The main innovations and improvements of the proposed next generation DT-centric EMS are discussed within this section are summarised in Tab. 4.4. These facilitate the operation and management of future power systems.

Tab. 4.4: Summary of main innovations of the proposed modular DT-centric EMS

Feature	Conventional EMS	DT-centric EMS
Architecture	Monolithic	Modular
Process IT	Server-client based approach	Middleware approach (transaction manager)
Data	Distributed, unconsolidated data, data access by single application	Central database, consolidated data, universal data access
Model base	Steady state model	Steady state and dynamic model
Modelling engine	Solver for algebraic models	Solver for differential-algebraic models
State estimation	Steady state	Steady state and dynamic state estimation
Model validation	Bad data detection for state estimation	Model validation and parameter estimation function
Process data	SCADA	SCADA and WAMS
Observability	Continuously updated stationary snapshots	Both continuously updated stationary snapshots and real time dynamics from WAMS
Decision support	Decision support based on the examination of contingencies and the manual investigation of possible solution options	Enabled for look ahead calculations, and automated investigation of preventive and curative measures based on a validated high-fidelity model
Trust	Limited trust and consideration of "worst case reserves"	High trust through validated model enables the utilisation of unexploited reserves
Advanced assistant functions	Not available	Novel applications for partially automated control up to temporary autonomous control become possible

5 Proof of Concept

As discussed in subsection 4.5, a validation of the concept presented in the thesis in a complete manner as done for industrial acceptance tests is hard to realise. To demonstrate the feasibility of the fundamental functions of the proposed DT-centred EMS, a "proof of concept" is carried out instead. For this purpose, an appropriate benchmark power system model has been implemented, and the DT as described before in subsection 3.1 is instantiated to prove the general working principle. The equipment models described in subsection 5.1, are implemented in the composite power system benchmark model, which is described and validated in subsection 5.2. To maintain an accurate model, the MHE parameter estimation method is verified in subsection 5.3. Synthetic PMU data streams are therefore used to verify and validate the dynamic model. As DSA studies play an increasing role in online power systems operation (see subsection 2.3.3), the proof of concept is dedicated to this task. The functionality and benefits of selected features of the proposed DT-centric EMS are demonstrated by numerical case studies in subsection 5.4.

5.1 Time-Domain Model Components

The time-domain models applied in the composite power system model are briefly introduced in the following subsections. Detailed model equations and corresponding parameters are documented in the appendix A.3 and A.5 respectively.

5.1.1 Synchronous Machine Model

A synchronous machine is a dynamic system in which electromagnetic and electromechanical processes take place and interact. Depending on the degree of detail i.e., the model order, the number of rotor coils on the direct axis (d-axis) and the quadrature axis (q-axis) and corresponding state variables vary from one to six [357]. The classical synchronous machine model for example, neglects all state variables of the rotor coils. Thus the field voltage and the corresponding rotor flux linkage is assumed to be constant [347, 358]. Due to its simplicity and computational efficiency, it is often applied to screen for transient stability limits [359], or to represent peripheral power plants in very large, interconnected power systems. However, investigations on power system stability have shown, that stability mechanisms are only correctly exhibited when synchronous machine models of fourth or higher orders are applied [202, 360]. For studies involving large perturbations (severe faults, such as 3-phase short circuits) turbo-generators should be modelled as sixth-order models [357]. Due to advantages in computing time, fourth-order models are often preferred for studies involving large power system models. While higher order models promise better results, the determination of their parameter sets requires more attention [358, 361].

Power system dynamics are mostly influenced by synchronous generators and their associated regulators and controllers, such as the turbine governor, the automatic voltage regulator (AVR) comprising the generator's excitation system, and power system stabilisers in some cases. A simplified block diagram, comprising the synchronous machine and its controllers is illustrated in Fig. 5.1.

The angular difference between the reference voltage u_{ref} depending on the excitation of the generator and the generator terminal voltage u_t, is given by the rotor angle δ. The turbine's mechanical power p_t and the excitation voltage v_{ex} are the synchronous machine input variables. A detailed formulation of the synchronous machine time-domain model, machine controller models and typical parameters applied within the numerical simulations in this thesis are documented in the appendix A.3.1.

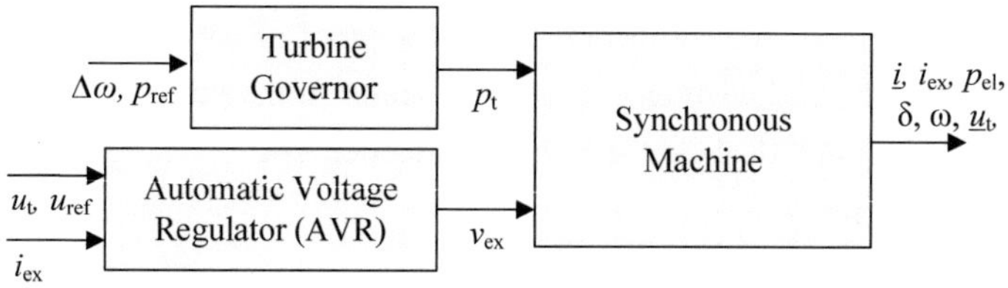

Fig. 5.1.: Simplified representation of the synchronous machine model with controllers

5.1.2 Voltage Source Converter

Today, voltage source converters (VSC) are regularly deployed as operating resources in the power system. They enable the exchange of energy between the alternating current (AC) grid and the direct current (DC) grid. Depending on the integrated controllers and control modes, it can directly influence the dynamics and quasi-stationary operating states of both AC and DC grid. The modelling of VSC-based high voltage DC (HVDC) transmission systems is extensively described in the literature (e.g. in [341, 362–367]) and is therefore only fundamentally addressed here. A CIGRÉ Type 6 (as described in [362]) is applied to represent the HVDC converters. Thus, the VSCs are modelled by controlled current and voltage sources in the phasor domain. The modelling approach ensures a sufficiently accurate representation of the described interactions between HVAC and HVDC transmission systems, and appropriate computational speed [362].

VSC-HVDC converters consist of a cascaded control structure, comprising an inner control scheme, i.e., the current control loop and outer controllers which control the VSC setpoints and determine the converter characteristics. The controllers themselves are usually realised by PI controllers comprising an integral and a proportional gain [362].

Normally, vendors in the industry do not disseminate detailed HVDC models, due to intellectual property issues [368]. Thus, a generic model derived from the literature has been implemented. The structure of the model follows the descriptions in [341, 365]. The definition of the current directions and the power balance correspond accordingly. Currents leaving the inverter on the AC side are counted positively, and those leaving the inverter on the DC side are counted negatively.

A converter station is illustrated in Fig. 5.2. The considered elements of the AC and DC side are shown as well as the philosophy of the separation of the inverter model and the DC grid. As shown in Fig. 5.2, the converter is modelled as a controllable current source on the AC side. Thus, the converter dynamics are influenced by the AC components of the VSC (i.e., transformer and filter characteristics). Since harmonics are not represented by the chosen phasor simulation approach, associated filter shunts can be neglected in the model.

Converter Control Loop

The block diagram of the converter control scheme is illustrated in Fig. 5.3. It comprises the inner and outer control loop, and the converter itself. The measured quantities are decomposed into their d- and q-components, to be controlled independently in the inner control loop. The inner control loop regulates the converter currents, which are then forwarded to the lowest converter control level, the valve control. As the valve or firing control is operating in the µs range, it is not considered in the described model. The outer control loop provides the reference signals i_d and i_q for the inner control of the converter and has the purpose to implement the higher-level converter control modes. By decoupling the d- and q-axes of the VSC through the inner control loop, it is possible to control the active and reactive power components fed into the AC grid independently. This is possible by adjustment of the reference values $i_{d,ref}$ and $i_{q,ref}$. Several control concepts are implemented in the model. These comprise active and reactive power flow control (mode P, mode Q), DC power flow control (mode P_{DC}), as well as AC voltage control (mode u_{AC}) and DC voltage control (mode u_{DC}). The controller transfer functions in the block diagrams illustrated in Fig. 5.3 are expressed in the frequency domain and are therefore denoted by the Laplace operator s. Further assumptions, the VSC model equations, control modes, controller equations and parameter sets are described in the appendix A.3.2.

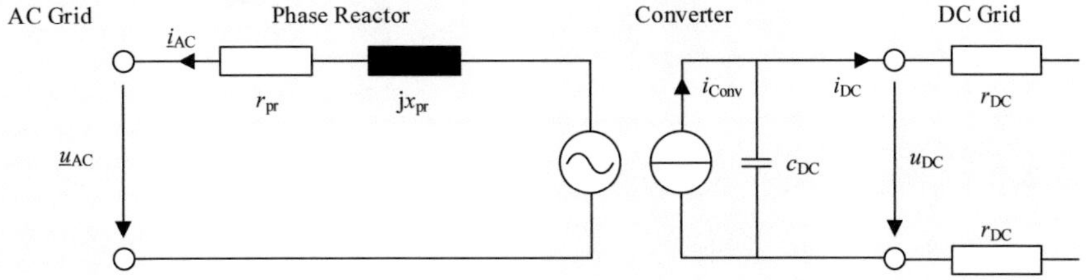

Fig. 5.2: Simplified single-line diagram of a voltage source converter model

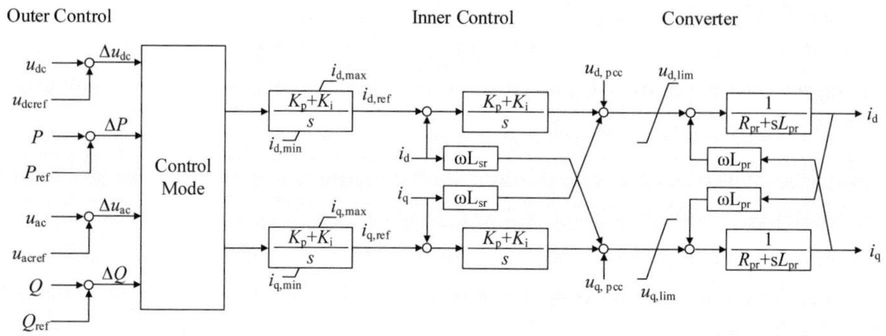

Fig. 5.3: Block diagram of the converter control scheme

5.1.3 Network Branch Model

The transmission line model is described by its series impedance $\underline{z}_{ij}$ and its charging susceptance b_{ij}. The standard π-branch model as shown in Fig. 5.4 is applied to approximate the electrical behaviour of AC transmission lines. Although a lumped parameter model can approximate the actual behaviour well, for modelling long lines a different expression considering the travelling wave effect and frequency dependent parameters, or a distributed parameter model is recommended to rise accuracy [5, 338]. Furthermore, the simplified transmission line model, does not consider asymmetry and mutual coupling between the transmission line wires. In the phasor domain, where the frequency is assumed to be close to the nominal system frequency, the simplifications of a lumped parameter model are acceptable. The generic network branch model allows furthermore to represent transformers, by a combination of a π-circuit as shown in Fig. 5.4 and an additional ideal transformer, which allows to model tap changing and phase angle shifts by a complex transformation ratio factor [369].

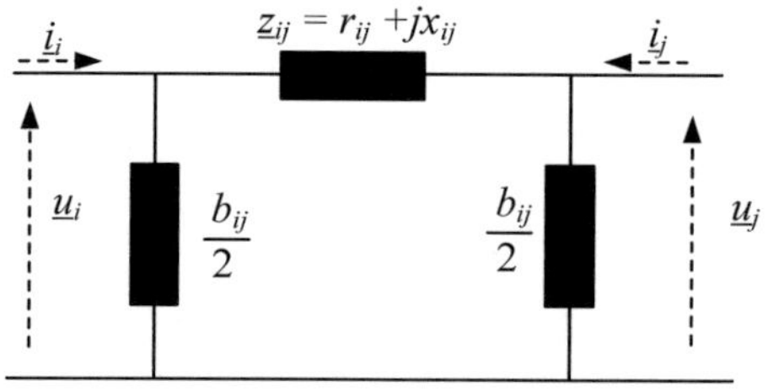

Fig. 5.4: Lumped parameter equivalent model of the transmission line

Assuming that the shunt conductance g_{ij} is negligible in comparison to the shunt susceptance b_{ij}, the circuit equations for the sending end i and receiving end j for a transmission line are related as follows:

$$\begin{bmatrix} \underline{i}_i \\ \underline{i}_j \end{bmatrix} = \begin{bmatrix} \underline{z}_{ij}^{-1} + b_{ij}/2 & -\underline{z}_{ij}^{-1} \\ -\underline{z}_{ij}^{-1} & \underline{z}_{ij}^{-1} + b_{ij}/2 \end{bmatrix} \cdot \begin{bmatrix} \underline{u}_i \\ \underline{u}_j \end{bmatrix} \tag{5.1}$$

Several techniques have been described in the literature to estimate transmission line model parameters from measurements (e.g., by applying regression [370], least-squares [371] or Kalman filtering [372]). Applying the branch model as shown in Fig. 5.4 (see subsection 5.1.3), the following circuit equations can be derived:

$$0 = \underline{i}_i - \underline{u}_i \cdot \frac{b_{ij}}{2} + \underline{i}_j - \underline{u}_j \cdot \frac{b_{ij}}{2} \tag{5.2}$$

$$0 = \underline{u}_i - \underline{z}_{ij} \cdot (\underline{i}_i - \underline{u}_i \cdot \frac{b_{ij}}{2}) - \underline{u}_j \tag{5.3}$$

Assuming that low noise phasor values of voltages $\underline{u}_i$, $\underline{u}_j$ and currents $\underline{i}_i$ and $\underline{i}_j$ are available at the sending and the receiving end of the transmission line, equations (5.2) and (5.3) can be substituted into equation (5.4) and (5.5) to derive $\underline{z}_{ij}$ and b_{ij} directly from measurements. Estimation techniques applying the Kalman filter deliver robust results under noisy measurement conditions. This is of special importance when accurate estimates of the shunt susceptance are required, which is typically about four to five magnitudes smaller than the series impedance [6].

$$\underline{z}_{ij} = \frac{\underline{u}_i^2 - \underline{u}_j^2}{\underline{u}_j \underline{i}_i - \underline{u}_i \underline{i}_j} \tag{5.4}$$

$$b_{ij} = 2 \frac{\underline{i}_i + \underline{i}_j}{\underline{u}_i + \underline{u}_j} \tag{5.5}$$

5.1.4 Modelling of Loads and Distributed Generation

A load in the context of grid calculations of higher voltage levels is an aggregated power consumption caused by several consumers at the point of electrical coupling. Consequently, it is challenging to model the load exactly because the aggregated load combines different types of consumers and is therefore difficult to characterise. Since loads can have a significant impact on the system voltage and dynamic stability, detailed information about the load composition is required, to model load dynamics. Load modelling approaches can be divided into component-based [373] and measurement-based [374, 375] approaches (see also [376, 377]). Thereby, the exponential dynamic load model described in [374], is a widely accepted model. It is also known as a composite ZIP model (where the three components determine the load model characteristics (Z = constant impedance, I = constant current, and P = constant power).

Although it is restrictive to assume constant impedance loads, most utilities worldwide use constant load models for power system stability studies [378], [379]. Constant loads are represented by an impedance to ground, i.e., by a shunt element $\underline{Y}_{i,\mathrm{load}}$ at the respective *i*-th bus in the admittance matrix $\underline{\boldsymbol{Y}}_{\mathrm{bus}}$ as described by equation (5.6). Although, this simplification leads to a voltage sensitive load equivalent, it is valid for studies addressing dynamic stability, that consider a short time frame following a disturbance [347, 380].

$$\underline{Y}_{i,\mathrm{load}} = \left(\frac{p_{\mathrm{load},i}}{u_i^2}\right) + j\left(\frac{q_{\mathrm{load},i}}{u_i^2}\right) = g_i + jb_i \tag{5.6}$$

The dynamic response of converter interfaced generation is dominated by their control algorithms, involving detailed vendor knowledge. Thus, generic dynamic models considering converter interfaced generation are hard to define. The authors in [381] recommend to benchmark phasor based models with detailed EMT models, to ensure that the controls are represented correctly. Although dynamic models are advantageous to gain insight into the real system behaviour, this may not always be possible since the vendors may not enclose these models to their products in the required format. Equal to load models, the simplest way to consider distributed generation is to build a steady-state model represented by a complex shunt element (i.e., equally to a negative load). This approach is still common worldwide and applied by utilities and system operators [378], and is therefore also applied to validate the approach within this work.

5.1.5 Initialisation of the Interconnected Power System Model

To build the interconnected power system model, the dynamic component models formulated as differential equations need to be interfaced with the power system represented by algebraic equations (see subsection 4.3.1). To solve the model equations, it is necessary to express all system variables in a common reference frame (see also appendix A.2). The interconnected transmission network as expressed by equation (5.8), is represented by buses, branches and static loads, which are considered in the nodal admittance matrix $\underline{\boldsymbol{Y}}_{\mathrm{bus}}$ by admittances (see also subsection 5.1.4). Assuming steady-state conditions, dynamic equipment models, such as generators, dynamic loads, and converters, are converted to Norton equivalents, and assembled to the network matrix by addition. This allows to compose the augmented admittance matrix $\underline{\boldsymbol{Y}}_{\mathrm{aug}}$ (5.7). By solving equation (5.8) the bus voltages $\underline{\boldsymbol{U}}_i$ at the *i*-th bus are subsequently determined.

$$\underline{\boldsymbol{Y}}_{\mathrm{aug}} = \underline{\boldsymbol{Y}}_{\mathrm{bus}} + \underline{\boldsymbol{Y}}_{\mathrm{Load}} + \underline{\boldsymbol{Y}}_{\mathrm{Gen}} \tag{5.7}$$

$$\underline{\boldsymbol{I}} = \underline{\boldsymbol{Y}}\,\underline{\boldsymbol{U}}\text{, or respectively } \begin{bmatrix} \underline{\boldsymbol{I}}_1 \\ \underline{\boldsymbol{I}}_2 \\ \vdots \\ \underline{\boldsymbol{I}}_i \end{bmatrix} = \begin{bmatrix} \underline{\boldsymbol{Y}}_{11} & \underline{\boldsymbol{Y}}_{12} & \cdots & \underline{\boldsymbol{Y}}_{1i} \\ \underline{\boldsymbol{Y}}_{21} & \underline{\boldsymbol{Y}}_{22} & \cdots & \underline{\boldsymbol{Y}}_{2i} \\ \vdots & \vdots & \ddots & \vdots \\ \underline{\boldsymbol{Y}}_{i1} & \underline{\boldsymbol{Y}}_{i1} & \cdots & \underline{\boldsymbol{Y}}_{ii} \end{bmatrix} \begin{bmatrix} \underline{\boldsymbol{U}}_1 \\ \underline{\boldsymbol{U}}_2 \\ \vdots \\ \underline{\boldsymbol{U}}_i \end{bmatrix} \tag{5.8}$$

Here, the vector $\underline{\boldsymbol{I}}_i$ represents all current injections or extractions by equipment such as generators, converters, and dynamic loads at the i-th bus. The DAE of the synchronous machines, exciters and governors, dynamic loads or VSC models, need to be initialised before an integration rule can be applied to solve the composite DAE model in the time-domain. To avoid long initialization times, the time-domain simulation is often started from a steady-state operating condition, which can be derived by state estimation or a power flow calculation. To obtain the value of the state variables of the DAE system at simulation time $t = 0$ s, the initial conditions are calculated from a steady state power flow result, and steady state conditions are validated by proofing the steady state criterion ($\dot{x} \approx 0$ i.e., $\dot{x} \leq 10^{-6}$). The values of the initial state variables x are obtained by calculation from the algebraic variables z, and the parameter set p. If the state variables are known (e.g., from a previous simulation or a DSE routine), the initialisation of the model simplifies accordingly. Since a wrong or invalid model initialisation leads to wrong simulation results [71], during the initialisation, screening for inconsistencies is mandatory to detect physically impermissible operating conditions, as well as state variables close to or exceeding saturation limits.

5.2 Description of the Benchmark Model

The benchmark model applied for validation of the concept is a modified version of the Cigré test system described in [382]. It has been intentionally developed to study the influence of embedded HVDC transmission systems on AC system performance and system security [382]. The model design therefore is especially useful to investigate dynamic phenomena, but is partly lacking realistic assumptions (e.g., long non-compensated AC transmission lines). The advantage of the model lies in its robustness to disturbances and in its relatively simple structure, which enables a straightforward validation of the model and its components. Furthermore, the model comprises all typical power system elements, to provide a suitable benchmark environment. It contains seven power plants, two HVDC converter stations, 24 busbars and 36 branches (including machine transformers). It can be divided into the two zones "North" and "South", which are interconnected by five AC transmission lines and one HVDC link. The base power flow scenario the northern zone is assigned to have a surplus of generation power, leading to a transit to the southern zone. The network model and the parameters of the component models are documented in appendix A.5. The topology of the base case is illustrated in Fig. 5.5.

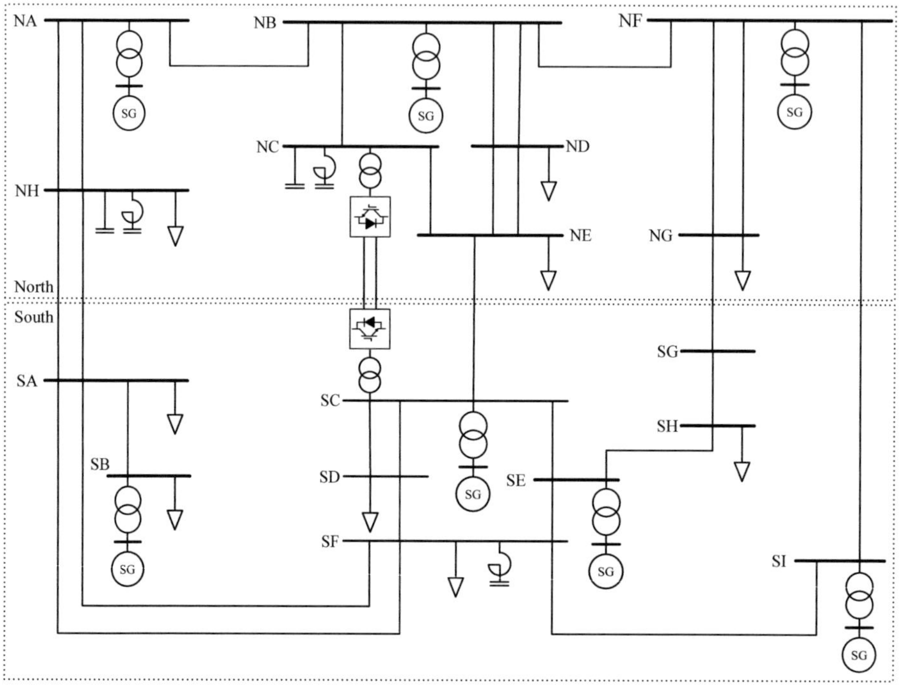

Fig. 5.5: Overview of the benchmark model topology in the base scenario

5.2.1 Benchmark Model Validation

As outlined before in subsection 2.7, one major obstacle to deploy advanced EMS applications is a lack of validated power system models and a reliable data basis. In consequence, any simulation conducted to support decision making in power system operation, relies on a model that can adequately replicate dynamic responses of the observed power system [190]. Therefore, a model validation procedure is of critical importance to ensure the required trust and confidence in the results. To validate the mirror model, the model response following an event is compared with the incoming synthetic telemetry data, created by the process simulation executed in DIgSILENT Power Factory. Therefore, relevant differential and algebraic states of the simulation are compared directly against each other (see subsection 5.2.2). The purpose of the benchmark model validation is to create a reliable basis to determine the accuracy of the mirror model. This ensures that the original system dynamics are known, and the deviation from the process simulation in case that the model parameters are falsified can be evaluated. All time-domain simulations in the mirror model are performed using the *Modified Euler* integration method (see appendix A.2.3).

5.2.2 Validation Results

The model response of the benchmark power system can only be accurate, when the underlying component models are correctly implemented. The validation of the component models has been part of the model development process (see also subsection 5.2.1). Considering this, the entire benchmark model itself is validated within this section, instead of validating each component model separately.

The mirror model accuracy must be high under diverse operating conditions. Therefore, several case studies and scenarios as described in Tab. 5.1 and Tab. 5.2 are defined. The cases in Tab. 5.1 represent different events as described in the table. The scenarios given in Tab. 5.2 refer to different load profiles, related to the base case (see appendix A.5). By a superposition of the cases and scenarios, 15 cases are created to evaluate the deviation between process and mirror model (see Tab. 5.3). To keep the system voltage for the initial conditions of scenarios B and C within normal range of operation, the shunt reactors and capacitors have been adapted accordingly to compensate excessive reactive power.

Tab. 5.1: Case study description

Case	Event	Description
0	No event	Simulation in steady state without event
1	Load step	Active and reactive power load step at bus NG to 50 %
2	Short circuit followed by transmission line trip	Short circuit event at bus SG at t = 1 s, trip of transmission line NG-SG at t = 1.1 s
3	Line Outage	Trip of transmission line NC-NE at t =1 s
4	Generation unit outage	Loss of generation unit at terminal SE at t =1 s
5	VSC setpoint adaption	Active power setpoint change of HVDC-VSC 1 at terminal NC from 400 MW to 200 MW

Tab. 5.2: Scenario description

Scenario	Scenario name	Description
A	High system load	Load profile according to the base case as given in the appendix A.5
B	Medium system load	Load and generation profile 50 % of base case A, Shunt reactors and capacitors adapted according to voltage band requirements
C	Low system load	Load and generation profile 25 % of base case A, Shunt reactors and capacitors adapted according to voltage band requirements

A simple metric to measure the difference between discrete time step aligned time series is to apply the Euclidean distance (ED) [383]. It represents the absolute distances between the observable time discrete process variables $\hat{y}_i^{\mathrm{m}}$ and the corresponding mirrored state variables $\hat{y}_i^{\mathrm{DT}}$ (5.9).

$$\mathrm{ED} = \sqrt{(\hat{y}_i^{\mathrm{DT}} - \hat{y}_i^{\mathrm{m}})^2} \tag{5.9}$$

Hence, the ED is easy to interpret and to evaluate. The accuracy of the modelling engine is validated by analysing the mean ED of selected time-discrete state variables in the process simulation and mirror model. The mean value of the ED is calculated by summing up the ED for the discrete samples in the time series and dividing the sum the total number of samples. The evaluation is done for an accurately tuned mirror model (i.e., an identical parametric setup) in comparison to the process simulation. For all conducted case studies, the highest resulting mean ED of selected state variable is given in Tab. 5.3. The corresponding simulation results and the ED between process simulation (dashed lines) and mirror simulation results (solid lines) are illustrated in Fig. 5.6 to Fig. 5.8.

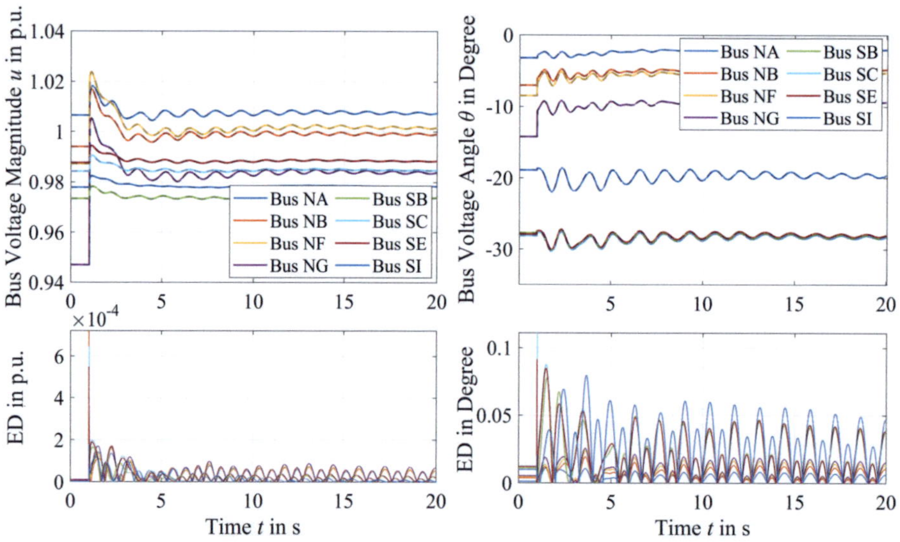

Fig. 5.6: Euclidean distance between process simulation (dashed lines) and mirror simulation results (solid lines) of the voltage magnitude u in p.u. and voltage angle θ in degrees at 400-kV power plant nodes for study case 1A

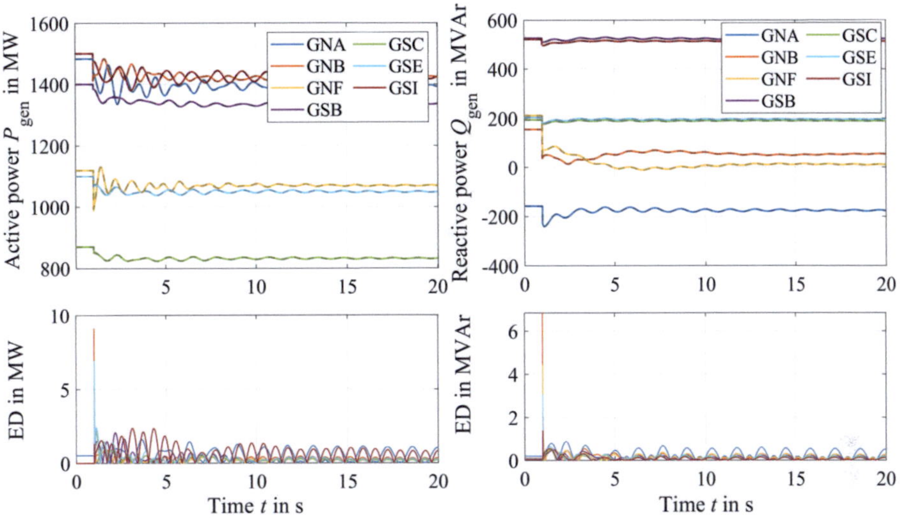

Fig. 5.7: Euclidean distance between process simulation (dashed lines) and mirror simulation results (solid lines) of the active power P in MW and reactive power Q in MVAr at the generator terminals for study case 1A

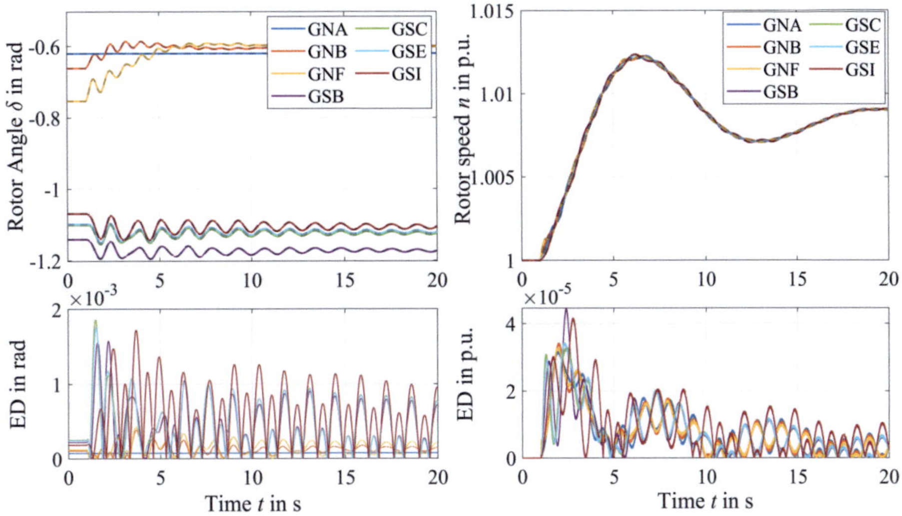

Fig. 5.8: Euclidean distance between process simulation (dashed lines) and mirror simulation results (solid lines) of the synchronous machine rotor speed n in p.u. and rotor angle δ in radians for study case 1A

The example trajectories of the variables illustrated in Fig. 5.6 to Fig. 5.8 show, that the dynamic mirror model can accurately reflect the process simulation, executed in DIgSILENT PowerFactory. The ED reaches its maximum shortly after the event execution and decreases with time. Spikes in the ED directly after the event execution result from the implemented numerical integration routines. These approximate the differential states differently in the discrete time steps directly following the event. Note that the rotor angle of synchronous generator GNA (Fig. 5.8) represents the angle reference and therefore remains constant in the chosen illustration.

To indicate the model accuracy, several algebraic and state variables are considered for model evaluation in Tab. 5.3, where the mean ED for these is summarised. They comprise the ac-voltage magnitude u_{ac}, the corresponding voltage angle θ, the synchronous machine rotor speed n, the active and reactive power P_{gen}, Q_{gen}, the VSC-HVDC dc-voltage u_{dc}, and the active and reactive power P_{VSC}, Q_{VSC}.

Comparing the mean ED of the variables for all scenarios given in Tab. 5.3, the order of magnitudes is equally low. Thus, no systematic error becomes evident. Furthermore, it can be concluded, that events causing a larger state transition, cause larger model deviations. The highest relative deviation between the models occurs in scenarios where the VSC-HVDC setpoint changes. Since the VSC-HVDC model implemented in the mirror model is simplified, the process simulation and the mirror model do not exactly match and cause a slightly higher deviation.

5.2.3 Measure for Model Accuracy

As the algebraic and dynamic state variables in the mirror model are resulting into the complex voltage phasor (see model equations in the appendix A.3.1), low voltage phasor errors between the mirror model and the observed voltage phasors in the power system indicate, that the model states are accurate, and the underlying model parameter set is appropriate. In consequence, the voltage vector error is a well-suited indicator to validate the fit between mirror model states and observed voltage phasors.

To evaluate the model accuracy in real-world applications, the total vector error (TVE) can be applied. It is usually applied to indicate the quality of the phasor estimated by a PMU. It expresses the difference between the estimated phasor and the true phasor value [211]. Following the IEEE standard [211], the TVE is expressed as per unit regarding the phasor value from the process (see also appendix A.4). Thus, a TVE of 0 p.u. represents the best possible fit between PMU phasor streams and the mirror model result. To apply the TVE as suggested, it is of major importance, that the timestamp of the incoming measurements is accurate, and the event location as well as the event time is recognisable from the measurements.

Tab. 5.3: Mean Euclidean distance for selected state variables (maximum value in bolt letters)

Case	Scenario	**Mean ED (t_{sim} = 20 s)**								
		$\|u_{ac}\|$	θ	n	P_{gen}	Q_{gen}	δ_{gen}	u_{dc}	P_{vsc}	Q_{vsc}
		p.u.	rad	p.u.	MW	MVAr	rad	p.u.	MW	MVAr
0	A	3.3e-06	1.2e-03	0.0e+0	5.1e-02	1.9e-02	2.6e-05	5.9e-07	8.5e-04	1.8e-06
	B	9.3e-07	1.0e-03	0.0e+0	4.7e-02	5.5e-03	2.9e-05	5.4e-07	8.1e-04	1.1e-11
	C	1.9e-07	9.7e-04	0.0e+0	4.5e-02	8.5e-04	4.5e-05	6.2e-07	7.8e-04	2.0e-07
1	A	3.3e-06	1.2e-03	1.9e-07	5.0e-02	1.8e-02	2.6e-05	4.3e-07	7.0e-04	8.9e-05
	B	9.1e-07	1.0e-03	9.8e-08	4.6e-02	5.2e-03	2.9e-05	5.1e-07	7.3e-04	3.0e-05
	C	2.1e-07	9.7e-04	4.9e-08	4.4e-02	7.8e-04	4.5e-05	5.5e-07	7.4e-04	1.9e-05
2	A	1.1e-05	**4.5e-02**	1.0e-06	5.8e-02	5.1e-02	6.9e-05	3.1e-05	1.1e-03	1.2e-03
	B	3.8e-06	4.4e-02	3.9e-07	5.1e-02	1.4e-02	4.8e-05	**4.1e-05**	2.0e-03	**2.3e-02**
	C	2.3e-06	4.5e-02	2.2e-07	4.8e-02	1.3e-02	4.5e-05	3.9e-05	1.7e-03	6.2e-03
3	A	3.3e-06	1.2e-03	3.9e-08	5.1e-02	1.9e-02	2.6e-05	5.3e-07	8.1e-04	2.3e-05
	B	9.4e-07	1.1e-03	1.2e-07	4.8e-02	5.8e-03	2.9e-05	6.9e-07	9.1e-04	6.2e-05
	C	2.1e-07	1.1e-03	1.9e-07	4.7e-02	8.9e-04	4.5e-05	7.6e-07	9.4e-04	9.6e-05
4	A	6.9e-06	1.6e-03	5.0e-07	5.4e-02	3.5e-02	**7.2e-03**	6.1e-07	1.7e-03	1.7e-04
	B	1.4e-06	8.9e-04	6.0e-08	4.6e-02	8.0e-03	3.8e-03	6.9e-07	7.8e-04	2.1e-05
	C	2.5e-07	1.1e-02	3.6e-07	4.1e-02	1.5e-03	2.0e-03	8.9e-07	4.6e-04	1.7e-05
5	A	**2.7e-05**	1.1e-02	2.0e-05	**1.7e-01**	**1.4e-01**	2.0e-04	1.5e-05	**1.9e-02**	5.8e-05
	B	9.9e-06	1.5e-02	**2.1e-05**	1.8e-01	4.9e-02	2.1e-04	1.5e-05	1.8e-02	3.3e-05
	C	3.9e-06	1.9e-02	2.1e-05	1.8e-01	2.3e-02	2.1e-04	1.5e-05	1.7e-02	1.9e-05

In cases where a high time delay or a large timestamp error between incoming PMU measurements and simulation result exists, the TVE is not an appropriate measure for model accuracy anymore.

Assuming these conditions are met, the TVE can be applied as an indicator of the mirror model accuracy. The mean TVE at 400-kV buses is illustrated in Fig. 5.9 to Fig. 5.11. It illustrates the effect of detuned model parameters on the TVE for the case studies resulting from the superimposition of the scenarios and cases described in Tab. 5.1 and Tab. 5.2. The illustration in Fig. 5.9 to Fig. 5.11 summarises the resulting TVE for a variation of the synchronous machine model parameter set between 0 % (i.e., considering a valid parameter set) and ±5 % (considering a deviating parameter set within the range of 5 % above and below the valid parameter set). Thus, p 0 denotes a valid parameter set in Fig. 5.9 to Fig. 5.11, p-5, a parameter deviation of 5 % below and p+5 denotes a parameter deviation of 5 % above respectively. It should be noted that the synchronous machines are the dominant voltages sources in the benchmark model, and therefore have major impact on the complex voltage phasor. Note that for illustrative purpose, the case study results are sorted according to their TVE deviation and are not presented in numerical order in Fig. 5.9 to Fig. 5.11. The following conclusions can be drawn from the model parameter variation experiment:

- In steady state, i.e., when no event occurs, the mean TVE remains close to zero,
- For small parameters deviations (within the range of ±1 %), the mean TVE remains below 1 %,
- In cases where the phase angle difference is high or is not accurately determinable, e.g., under fault conditions (case study 2), or loss of a generation unit (case study 4), the mean TVE rises disproportionately with rising parameter deviations,
- As the amount of power fed into the grid after a VSC-HVDC setpoint change (case study 5) is the same for all scenarios the mean TVE rises in relation for scenarios with lower system load (see Tab. 5.2),
- For the investigated case studies applying the benchmark model, the TVE can provide a measure for model accuracy.

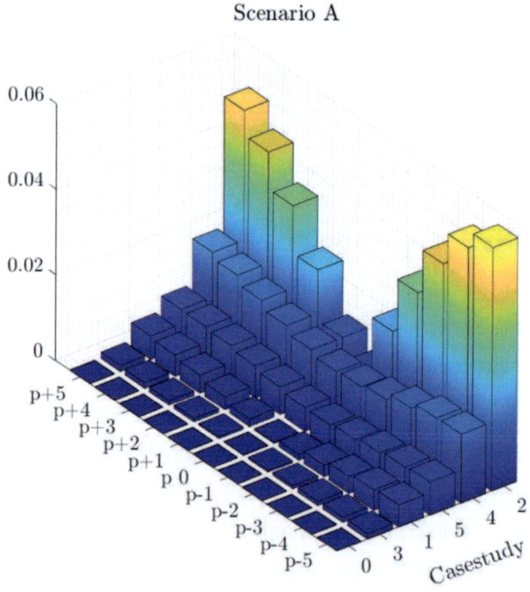

Fig. 5.9: Mean TVE in p.u. for varying synchronous machine parameters in scenario A and different case studies as described in Tab. 5.1 and Tab. 5.2

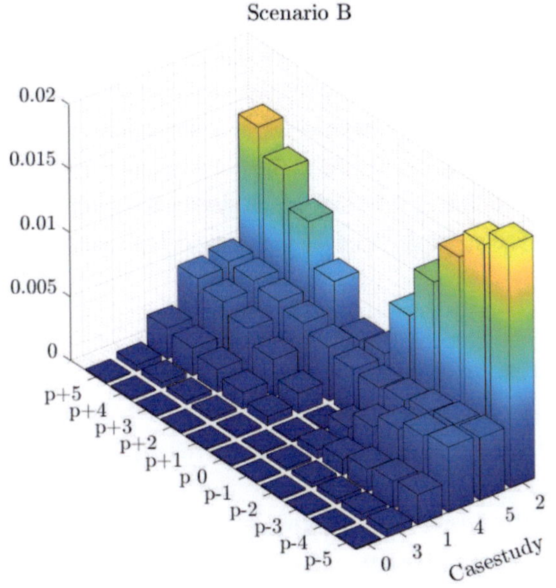

Fig. 5.10: Mean TVE in p.u. for varying synchronous machine parameters in scenario B and different case studies as described in Tab. 5.1 and Tab. 5.2

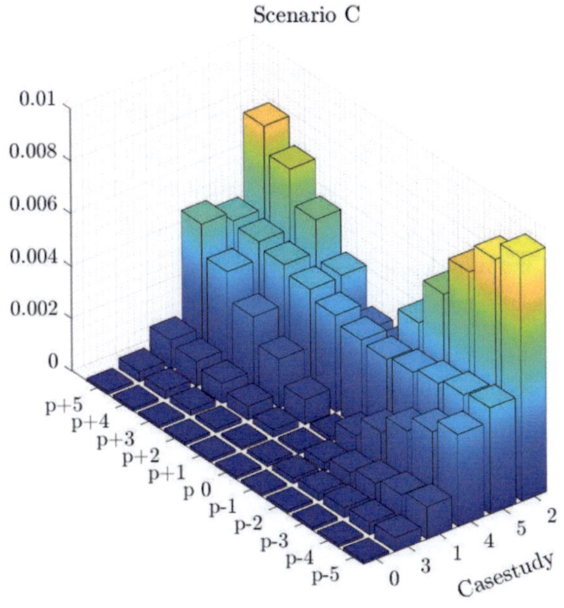

Fig. 5.11: Mean TVE in p.u. for varying synchronous machine parameters in scenario C and different case studies as described in Tab. 5.1 and Tab. 5.2

5.2.4 Conclusion on the Mirror Model Validation

The results given in subsection 5.2.2 illustrate that the ED between mirror model and process simulation is significantly larger during the dynamic transitions and for a short period afterwards. Thus, also the TVE decays as the derivatives of the differential states during the numerical integration get smaller. After the fast transient transitions have subsided, the deviations decay dynamically depending on the system damping and the extent of the excitatory event. It can be concluded that the mirror model can accurately reproduce the process simulation.

Referring to the modelling assumptions given in subsection 4.3.1, the coupling of differential states and algebraic states as well as the grid topology and branch parameters are expressed within the complex system voltage $\underline{u}_{\mathrm{ac}}$. This property has been proven by a numerical experiment by assessing the mean TVE for several superimposed scenarios. The correlation between the synchronous machine parameter accuracy and the mean TVE has been examined. For a TVE below 0.01 p.u., the model parameter set can be considered accurate for the investigated number of scenarios. Thus, the TVE is applicable to provide a measure to assess the model accuracy in both offline and online model validation routines.

5.3 Dynamic Model Parameter Estimation

Although most component model parameters are available in the form of data sheets, these are sometimes implemented wrong, and rarely checked afterwards. In rare cases it is possible to validate the model against measurements. To avoid the violation of validity boundaries of the model, resulting in invalid results or even in the worst case in unfortunate or devastating operational decisions, the ENTSO-E recommends a continuous validation and fine tuning of system models [71]. A suitable method for power system model tuning has been introduced in subsection 3.4.2. As an input the method requires PMU based WAMS data records, to fit the model response by parameter adaption according to the physical system response during operation.

The MHE technique applies a time horizon with N samples and minimises the distance between the given model response and the measurement data. To illustrate the efficiency of the parameter estimation technique, a sixth-order synchronous machine model is taken as an example to reconstruct the model parameters from the dynamic responses of the process simulation. The incoming measurement samples have a fixed rate and share the characteristics of discrete PMU samples (see also subsection 2.4.1).

To add further information to the estimation problem, the set of parameters being searched for is constrained by feasible physical bounds and apparent relations between synchronous, transient, and sub-transient dimensions [5, 361]. A typical range for synchronous machine models is given in Tab. 5.4. These can be considered as meaningful constraints for the parameter estimation procedure. From the given open-circuit and short-circuit machine parameters, the equivalent circuit resistance and reactance can be calculated [384, 385]. The parameter constraints are considered as lower $\underline{\boldsymbol{p}}$ or upper bounds $\bar{\boldsymbol{p}}$ to the parameter set $\boldsymbol{p}$ i.e., $\underline{\boldsymbol{p}} \leq \boldsymbol{p} \leq \bar{\boldsymbol{p}}$ by the MHE. In case that the optimality criterion for the observed model response is reached, and the parameter constraints are fulfilled, the model accuracy is rated as appropriate (see also subsection 5.2.1).

Tab. 5.4: Typical range of parameters for synchronous machine models

Parameter	Parameter name	Range of typical parameter values [5]
T_A	Acceleration time constant ($T_A = 2H$)	4.0 - 12 s
D	Damping constant	0 – 2
x_d	Synchronous reactance d-axis	0.6 – 2.3 p.u.
x_q	Synchronous reactance q-axis	0.4 – 2.3 p.u.
x'_d	Transient reactance d-axis	0.15 – 0.5 p.u.
x'_q	Transient reactance q-axis	0.3 – 1.0 p.u.
x''_d	Subtransient reactance d-axis	0.12 – 0.35 p.u.
x''_q	Subtransient reactance q-axis	0.12 – 0.45 p.u.
x_l	Stator leakage inductance	0.1 – 0.2 p.u.
r_{str}	Stator resistance	0.0015 – 0.02 p.u.
T'_{d0}	Transient open circuit time constant d-axis	1.5 – 10.0 s
T'_{q0}	Transient open circuit time constant q-axis	0.5 – 2.0 s
T''_{d0}	Subtransient open circuit time constant d-axis	0.01 – 0.05 s
T''_{q0}	Subtransient open circuit time constant q-axis	0.01 – 0.09 s

The general formulation of the MHE objective function given in subsection 3.4.2 can be reformulated considering, that the computation of the algebraic states $\boldsymbol{z}(t) = \boldsymbol{z}(\boldsymbol{x}(t))$ is done independent from the differential states, whenever these are uniquely determinable. Thus, the interrelations of the equations can be written in the form $\boldsymbol{z}(t) = \boldsymbol{A} \cdot \boldsymbol{b}(\boldsymbol{x}(t))$ with the constant matrix $\boldsymbol{A}$ and a time-varying vector $\boldsymbol{b}(\boldsymbol{x}(t))$. Thus, the algebraic states can be removed from (3.3)-(3.5), which then simplify to the ordinary differential equation system (5.10)-(5.11), with $\hat{\boldsymbol{f}}(\boldsymbol{x}, \boldsymbol{u}, \boldsymbol{p}) = \boldsymbol{f}(\boldsymbol{x}, \boldsymbol{A}\boldsymbol{b}(\boldsymbol{x}), \boldsymbol{u}, \boldsymbol{p})$ and $\hat{\boldsymbol{h}}$ analogous.

$$\dot{\boldsymbol{x}}(t) = \hat{\boldsymbol{f}}(\boldsymbol{x}(t), \boldsymbol{u}(t), \boldsymbol{p}) \tag{5.10}$$

$$\boldsymbol{y}(t) = \hat{\boldsymbol{h}}(\boldsymbol{x}(t), \boldsymbol{u}(t), \boldsymbol{p}) \tag{5.11}$$

This simplification leads to the objective function J given in (5.12).

$$J(\hat{\boldsymbol{x}},\hat{\boldsymbol{p}},\hat{\boldsymbol{u}},\hat{\boldsymbol{y}}^{\mathrm{m}})=\left\|\begin{matrix}\hat{\boldsymbol{x}}(k-N)\\ \hat{\boldsymbol{p}}(k-N)\end{matrix}-\chi^{k}\right\|_{S}^{2}+\sum_{n=k-N}^{k}\left\|\hat{\boldsymbol{h}}(\hat{\boldsymbol{x}}(n),\hat{\boldsymbol{u}}(n))-\hat{\boldsymbol{y}}^{\mathrm{m}}(n)\right\|_{R}^{2}$$
$$+\sum_{n=k-N}^{k-1}\left\|\begin{pmatrix}\hat{\boldsymbol{x}}(n+1)-\hat{\boldsymbol{x}}(n)-\Delta t\hat{\boldsymbol{f}}\left(\hat{\boldsymbol{x}}(n),\hat{\boldsymbol{u}}(n),\hat{\boldsymbol{p}}(n)\right)\\ \hat{\boldsymbol{p}}(n+1)-\hat{\boldsymbol{p}}(n)\end{pmatrix}\right\|_{Q}^{2} \tag{5.12}$$

The differential equations given in the appendix (A.39)-(A.44), and the algebraic equations (A.45)-(A.47) from appendix A.3.1 are taken as input model to the MHE. Together, these equations describe the system dynamics of the synchronous machine. The corresponding outputs $\hat{\boldsymbol{y}} = (n, \delta, i_{\mathrm{ex}}, \underline{i}_{\mathrm{t}}, \underline{u}_{\mathrm{t}}) : [0, \infty) \rightarrow \mathbb{R}^3, \mathbb{C}^2$ can be calculated by the equations (A.48)-(A.52) given in the appendix A.3. To estimate the parameters in addition to the differential states, these are treated as states with zero dynamics i.e., $\dot{\boldsymbol{p}}(t) = 0$ (see also subsection 3.4.2).

The machine rotor angle δ and the excitation current i_{ex} are not directly observable by the PMU based WAMS. Thus, only the electrical frequency, the current injected into the grid and the complex voltage can be taken as input $\hat{\boldsymbol{y}}^{\mathrm{m}}$ for the estimation algorithm. The electrical frequency as an input signal can be approximated by the machines rotor speed n [386], [387].

The synthetic measurements $\hat{\boldsymbol{y}}^{\mathrm{m}}$ generated by the process simulation are subjected to random Gaussian noise $\boldsymbol{w}$, i.e., $\hat{\boldsymbol{y}}^{\mathrm{m}}(n)=\hat{\boldsymbol{y}}(n)+\mathbf{w}(n)$ with zero mean ($\mu = 0$) and a constant variance ($\sigma = 1$ %) is intentionally added to evaluate the robustness of the MHE algorithm. Thus, in the experimental setup, the noisy time series data is validated against the model response to prove the robustness of the MHE under realistic conditions.

To investigate the impact of the horizon length N on the estimation accuracy, the MHE horizon length has been varied. For better comparability, the same random error $\boldsymbol{w}$ is applied on $\hat{\boldsymbol{y}}^{\mathrm{m}}(n)$ for the investigation. The MHE experiment is illustrated for a period of 200 ms in discrete time steps of 10 ms directly following a load step event subjected to the synchronous machine. In consequence, the internal states change as well as the observable measurands $\hat{\boldsymbol{y}}^{\mathrm{m}}$ at the machine terminal. The initial parameter set $\boldsymbol{p}$ is biased by an offset of 20 % and constrained by lower and upper bounds (see Tab. 5.4). The results of the MHE are shown in Fig. 5.13 to Fig. 5.16 and are summarised in Tab. 5.5. The results given in Tab. 5.5 show, that the parameters estimated for each investigated time horizon N are closely bound to the mean value μ. While most parameters are estimated very accurate, a few have a relative high deviation in comparison to the referenced original parameter value p_{ref}. According to the model equations given in the appendix A.3.1 (see also [385]), the leakage reactance x_{l} is only considered indirectly, resulting in a weak correlation to the original value p_{ref}. In consequence, the highest parameter deviation can be found for the leakage reactance x_{l}.

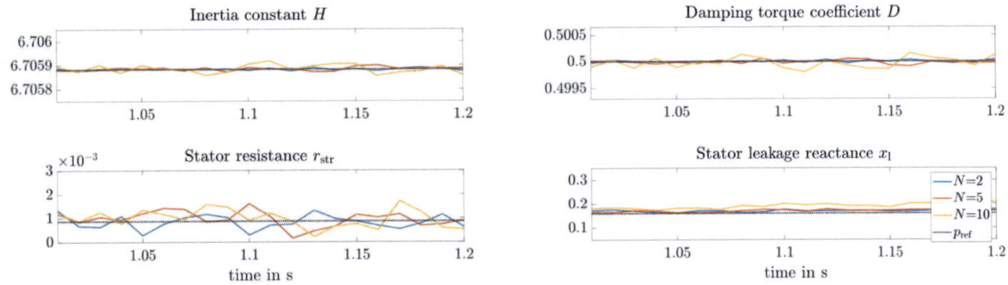

Fig. 5.12: MHE results for inertia constant, damping torque coefficient, and synchronous machine stator parameters

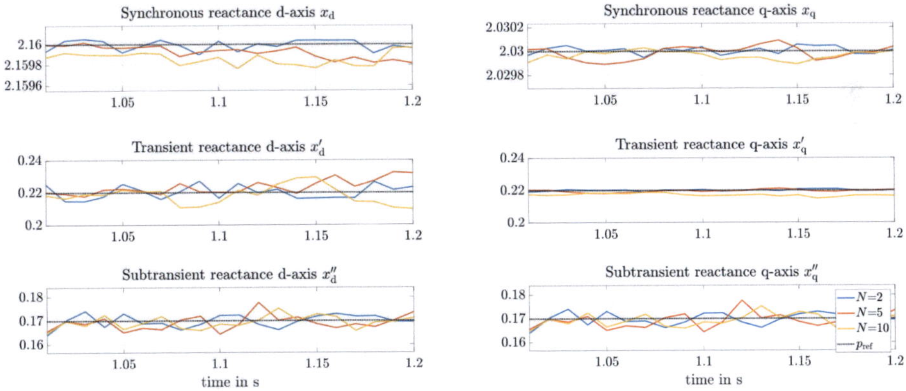

Fig. 5.13: MHE results for synchronous, transient, and sub-transient synchronous machine reactance parameters

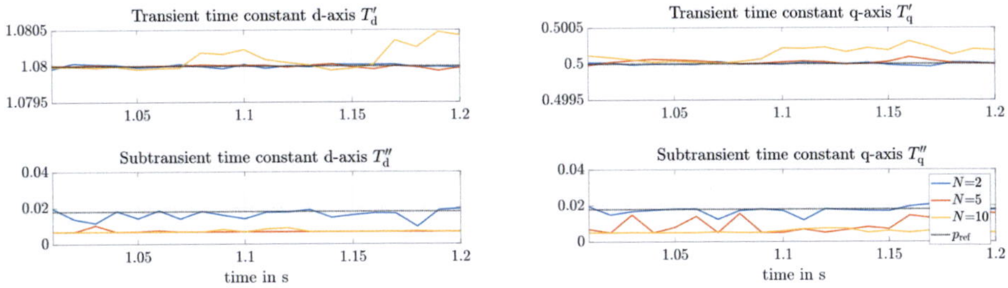

Fig. 5.14: MHE results for the synchronous machine time constants

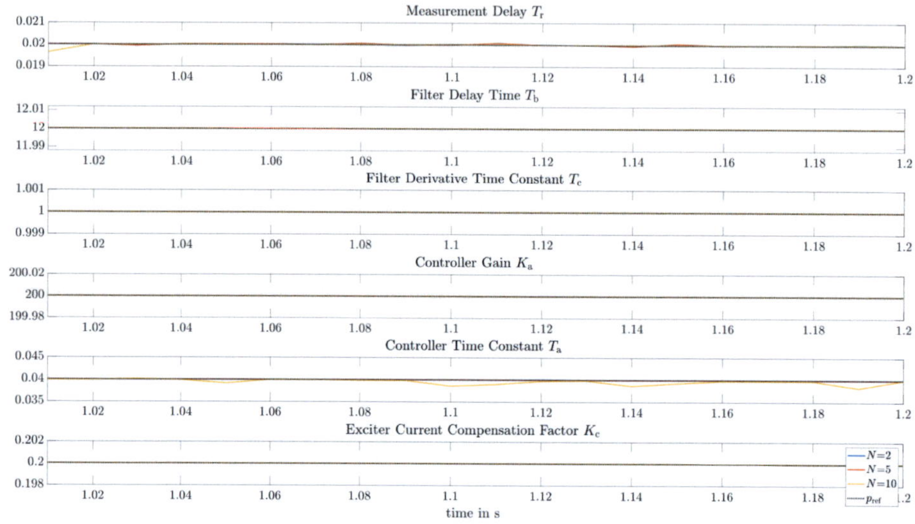

Fig. 5.15: MHE results for the IEEE AC4C excitation system

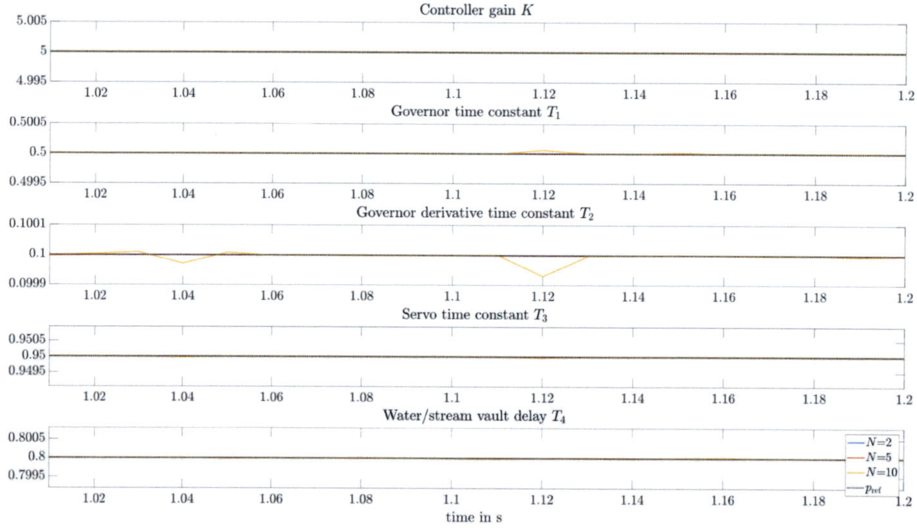

Fig. 5.16: MHE results for the governor parameters comparing horizon length N

Tab. 5.5: Original parameters, mean μ and standard deviation σ of the estimation results

Parameter		Original value p_{ref}	Estimated parameter value N = 2 μ	σ	N = 5 μ	σ	N = 10 μ	σ
Synchronous machine	H	6.7059	6.7059	4.29e-06	6.7059	7.07e-06	6.7059	1.795e-05
	D	0.5	0.5	2.11e-05	0.499	3.67e-05	0.4999	9.95e-05
	r_{str}	8.7e-4	8.28e-04	2.92e-04	9.65e-04	3.32e-04	9.87e-04	3.89e-04
	x_l	0.16	0.169	3.39e-03	0.17	4.6e-03	0.19	9.813e-03
	x_d	2.16	2.16	4.96e-05	2.159	6.26e-05	2.1599	6.212e-05
	x_q	2.03	2.03	3.06e-05	2.03	5.49e-05	2.03	3.639e-05
	x'_d	0.22	0.22	4.08e-03	0.223	4.29e-03	2.182e-	5.578e-03
	x'_q	0.22	0.22	5.02e-04	0.2196	7.70e-04	0.217	9.689e-04
	x''_d	0.17	0.1699	2.59e-03	0.1691	3.04e-03	0.1694	2.611e-03
	x''_q	0.17	0.1699	2.59e-03	0.1691	3.04e-03	0.1694	2.611e-03
	T'_d	1.08	1.08	2.23e-05	1.08	2.16e-05	1.08	1.63e-04
	T'_q	0.5	0.5	1.64e-05	0.5	2.71e-05	0.5	8.846e-05
	T''_d	0.018	0.0162	2.68e-03	0.00689	7.68e-04	0.00695	6.29e-04
	T''_q	0.018	0.0169	2.30e-03	0.00943	4.28e-03	0.00552	7.91e-04
Governor	K	5.0	5.0	8.83e-08	5.0	2.36e-07	5.0	2.292e-06
	T_1	0.5	0.5	4.92e-07	0.5	8.61e-07	5.0	1.401e-05
	T_2	0.1	0.1	1.75e-07	0.1	5.56e-07	0.999	1.681e-05
	T_3	0.95	0.95	6.65e-08	0.95	2.89e-07	0.95	9.452e-06
	T_4	0.8	0.8	3.64e-07	0.8	2.39e-06	0.8	1.067e-05
Exciter	T_r	0.02	0.02	1.32e-05	0.2	3.22e-05	0.1998	7.983e-05
	T_b	12	12.0	2.26e-08	12	7.72e-08	12	7.937e-07
	T_c	1	1.0	1.03e-07	1.0	6.97e-07	1.0	4.60e-06
	K_a	200	200	1.92e-06	200	1.08e-06	200	3.879e-05
	T_a	0.04	0.04	1.25e-05	0.3997	2.23e-05	0.395	5.351e-04
	K_c	0.2	0.2	2.37e-07	0.2	1.40e-07	0.2	4.469e-06

Conclusion

The experiment shows that observable system variables and captured dynamics contain meaningful information, which can be applied to estimate time-domain model parameters. The applied MHE method works well for DAE models such as the investigated sixth-order synchronous machine model including the connected feed forward controllers. Multiple parameters can be estimated simultaneously, although estimating a smaller number of parameters at once will result in higher accuracy. The intentionally added noise does not influence the estimation result significantly. Thus, it can be concluded that the MHE is robust to measurement noise. In contrary to the expectation that the accuracy increases with increasing horizon N, this could not be observed for all cases. Especially for the d-axis transient reactance x'_d and the leakage reactance x_l the parameter estimation results become less accurate although more information is available. In these cases, the underlying optimisation routine is no longer converging towards the optimality criterion resulting in higher accuracy. Nevertheless, the parameter deviation from the reference value is relatively small and acceptable. The MHE also achieves accurate results with small horizons (i.e., $N = 2$). Since the optimization problem is significantly reduced for smaller horizon lengths, the computation time decreases accordingly. This proves, that the MHE can be applied as a parameter estimation routine within the proposed DT-centric EMS framework and can be considered for real-time applications.

5.4 Illustrative Case Studies

The purpose of the proposed DT-centric EMS architecture is to enhance existing EMS applications (see subsection 2.3) and to enable novel automation functions for power system analysis and control. A selection of enabled functions is discussed and demonstrated by case studies within this subsection to illustrate the capability of the DT-centric EMS architecture. Furthermore, the concept is validated by providing a "proof of concept" by these dedicated case studies. Besides enhanced dynamic observability, the DT-centric EMS provides a sand-box instance to benchmark operational decisions as well as remedial action schemes. In addition, the gained trust in the validated model within the DT-centric EMS increases the confidence into the simulation results, and their applicability within decision support systems for online system operation (e.g., for dynamic security assessment).

5.4.1 Enhanced Observability

As discussed earlier, the DT framework has the purpose to enhance the power system observability in the control centre. As the classic SSE is based on power flow algorithms (see subsections 2.3.1 and 2.3.2), upcoming volatile situations in power system operation may result in non-convergence. In consequence all subsequent dependent EMS functions and modules are not working properly. A partial loss or full loss of observability endangers the security from the perspective of system operation. Thus, sudden changes in power generation (e.g., by cloud shade or wind gusts) can result in challenging situations from the perspective of power system operation.

Although there are many possible reasons for non-convergence of the power flow equation-based state estimation, the main reasons during real-time operation are mainly driven by bad data quality, data loss or delayed data packages as well as high gradients in state changes (e.g., by volatile wind or solar generation). These state changes can cause that the

Jacobian matrix required to solve the power flow equations becomes singular, i.e., non-invertible, resulting in non-convergence. It comprises the first-order partial derivatives of the power flow equations with respect to the state variables.

An advantage of the calculation in the time-domain is that the system matrix does not have to be inverted to solve the system equations. This allows to include the state change directly into the calculation and to observe the consequential state changes of all other state variables. To illustrate this, a case scenario is created by slightly altering the base scenario A of the Cigré benchmark system described in subsection 5.2. To create this specific situation in the benchmark model, the power consumption of static load at bus SA is reduced to 75 % and the static load at bus NG is set to 50 % of the original base case value. Furthermore, the power consumption of load at bus NH is set to $P = 1600$ MW and $Q = 376$ MVAr. The shunt capacitor at the same bus is set to $Q = 200$ MVAr to maintain the system voltage. This setting results in a high loading scenario in the north-western area in the modelled system. An additional sudden drop in power generation (e.g., by cloud shade or wind gusts) now leads to a high residual load at node NH.

By rising the residual load at node NH in form of a continuous power flow (CPF) study step wise, a so called "*nose curve*" can be derived. As illustrated in Fig. 5.17, the power flow converges up to a power consumption of approximately 1.28 p.u. of the referenced value at bus NH. In the illustrated CPF study, this corresponds to an incrementally increased active and reactive power consumption at bus NH from 1600 MW to 2048 MW and from 200 MVAr to 481 MVAr. Above this power consumption at bus NH, the system equations remain unsolved, and the CPF is stopped since the Newton-Raphson power flow algorithm does not converge anymore.

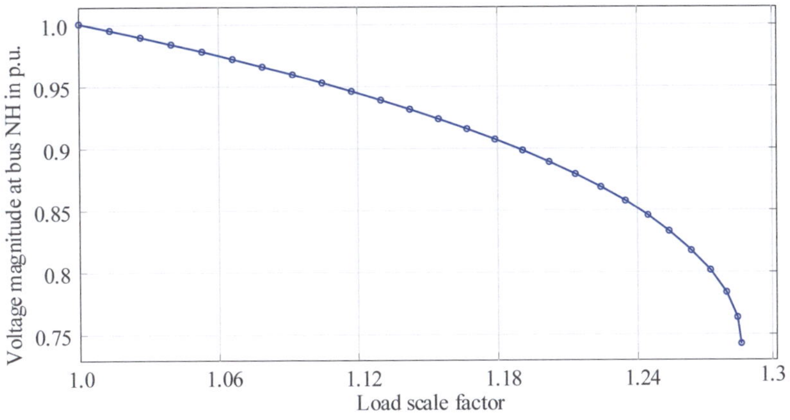

Fig. 5.17: "Nose curve" for bus NH obtained by a continuous power flow study

Starting at the same conditions within the time-domain model, the system equations remain solvable although the load at bus NH increases above the given scale factor of 1.28 p.u. The advantages of the time-domain model become obvious in this scenario. In comparison to the power flow calculations done for a CPF study, in the time-domain it is not necessary to invert the Jacobian matrix. Furthermore, the frequency and voltage controllers are considered in the time-domain model. In consequence, the system voltage can be retained by the controllers of the active power plants following the sudden drop in power generation. This is illustrated in Fig. 5.18. Here, the time-domain model is able to resolve load changes above the CPF limit (see dashed line in Fig. 5.18). Thus, the system remains observable. In addition, the load model characteristics (see subsection 5.1.4) become evident in Fig. 5.18. While the voltage gradient shows linear characteristics, the active and reactive power are varying in a nonlinear way i.e., follow a quadratic function. Therefore, a general conclusion can be drawn. By considering the state change directly in the time-domain model the consequential state changes of all other state variables can be observed. Thus, enhanced dynamic observability is given. Furthermore, the distance to the dynamic security limit can be determined (e.g., by considering the excitation limiters of all power plants in the observed area).

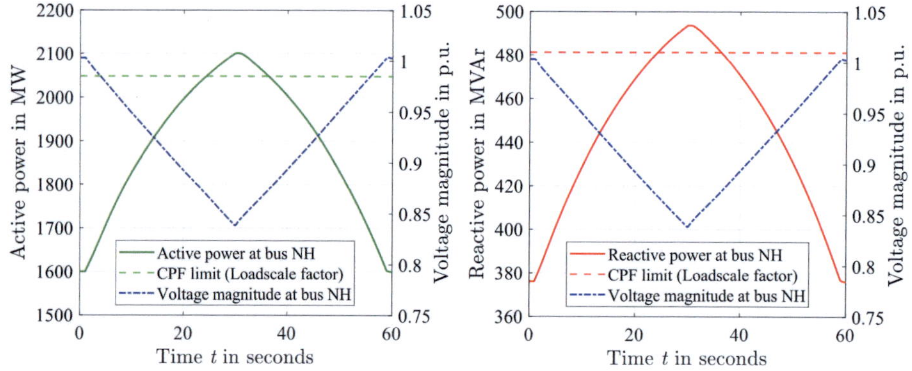

Fig. 5.18: Time-domain simulation results at bus NH

5.4.2 Online Model Validation

The tracking of sudden changes of the system load is also illustrated in Fig. 5.19 and Fig. 5.20. Both the active and the reactive power consumption are varying over time. To illustrate the impact of the model parameter accuracy, the average TVE for all 24 buses in the benchmark model is illustrated for the tuned system as well as for the detuned system. Here the parameters of the synchronous machine are subjected to an error of -5 % in comparison to the reference system simulated in PowerFactory. For the tuned system, the TVE decays close to zero after the transient transition ends, whereas for the detuned system model, the error remains around 0.22 % at the end of the time series. The maximum deviation is almost equal for both, but the mean error is significantly lower for the tuned system model.

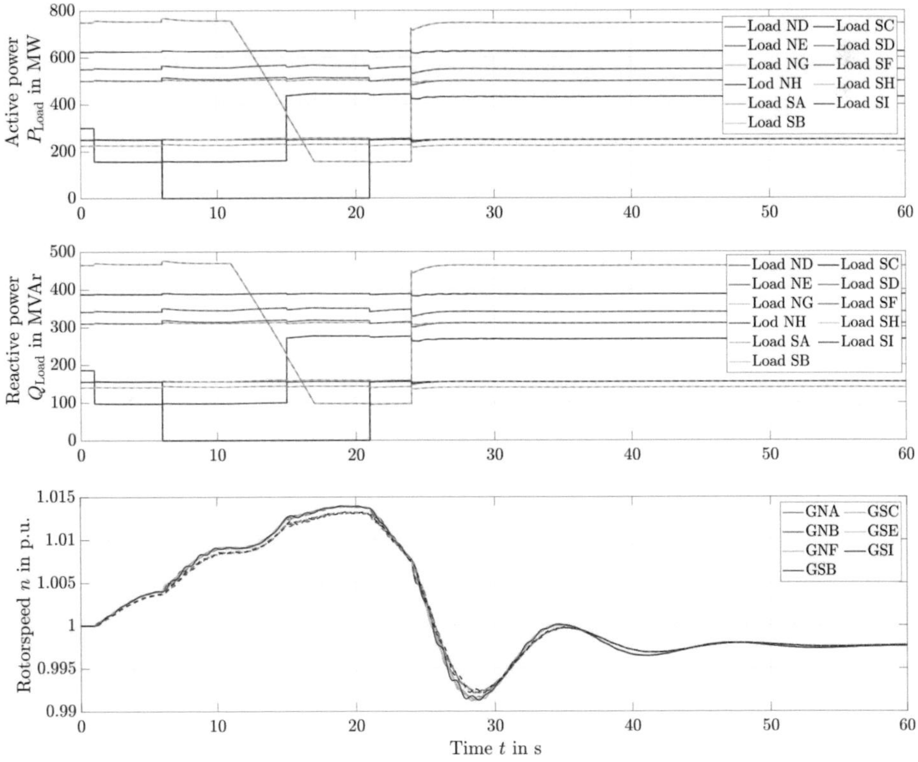

Fig. 5.19: Varying active power P_{Load} and reactive power Q_{Load} and corresponding generator rotor speed n for tuned (solid line) and detuned model (dashed line)

Conclusion

The case study given in subsection 5.4.1 shows the advantage of the dynamic power system model in comparison to the steady state power flow model. A well-known weakness of power flow calculations is that the Jacobian matrix required to solve the power flow equations becomes singular, i.e., non-invertible under certain conditions. Unfortunately, this can also occur during operation, when the point of operation is close to operational limits, or systems under high loading. Thus, a non-convergence of the power flow equations leads to a certain level of unobservability. In consequence, state estimation and power flow algorithms do not necessarily converge during such volatile situations. Furthermore, the investigations in subsection 5.4.2 have shown, that a tuned DAE model is able to reproduce changes in the observed process better. Thereby, enhanced observability is obtainable. It can be concluded that the proposed dynamic mirror based on DAE models can enhance observability. Furthermore, system states, which are not obtainable by conventional measurements become observable and applicable for analytic functions, or as a surrogate for missing data. This is of importance for implementation of advanced assistance systems, or cognitive systems which may require a large amount of training data (see also [282]).

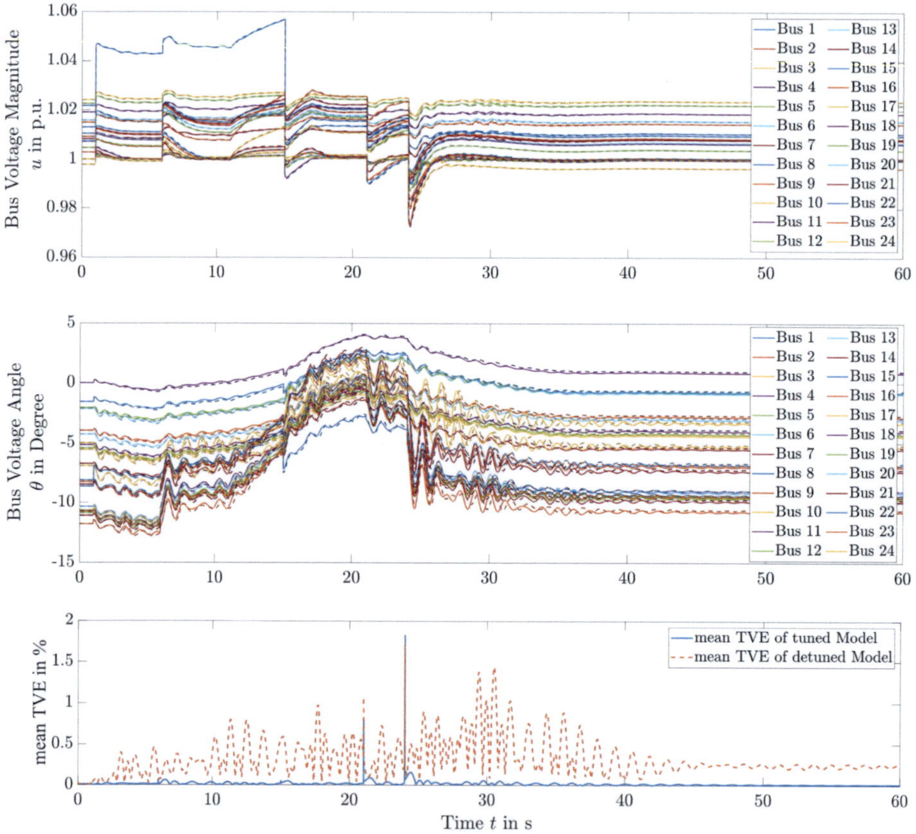

Fig. 5.20: Bus voltage magnitude u and corresponding voltage angle θ, resulting in the mean TVE for tuned (solid line) and detuned model (dashed line)

5.4.3 Reliable Dynamic Security Assessment

As previously described in subsection 2.3.3 the terms security and stability in the context of electrical energy systems relate to the behaviour, that the system returns to an equilibrium after the occurrence of a disturbance. Thus, the system regains a quasi-stationary state i.e., $\dot{x} \approx 0$, once the excitatory event has subsided. To assess this online, a DSA can be executed to investigate prospective operational scenarios. For most online DSA tools in transmission systems the following indicators and system variables are of interest (see also [71]):

- States and limit violations of electrical power system variables (voltages, currents, frequency),
- Equipment loading, and violated security margins (i.e., exhaustion of primary control reserve, reactive power reserve, or temporary admissible equipment loading),
- Minimal and maximal values reached during dynamic transitions,
- Excitation or tripping of protection devices or protection schemes,
- High frequency gradients, stress on turbines, or machine excitation limits.

The overall goal of DSA tools is to maximise the system resiliency to severe disturbances (see also subsection 2.2.2). This is done by coupling the power system model to the latest state estimation result derived from the real-time process, and screening for possible contingencies taking the actual and prospective future points of operation into account.

Obviously, a correctly parametrised and validated power system model is required to conduct meaningful DSA studies. However, an incorrect parametrisation may be hard to identify by manual model checks. Online DSA in practice is done for several possible scenarios, resulting in hundreds of assessed contingencies. Thus, an incorrectly parametrised component or controller model may impact a significant number of results. The effects of parameter inaccuracies cannot be generalised, as their impact depends on the operating point and is often influenced by non-linearity and may influence the system behaviour only in specific situations or operating conditions. This directly impacts the reliability of decisions made from the available information. Therefore, it is of crucial interest to keep the model base valid during online operation. A wrong parameter describing the generators saturation for example leads to an underestimated or overestimated short circuit current contribution. The frequency response in turn is very sensitive to the given inertia constant. To illustrate the impact on voltage stability, an example is chosen, where the exciter gain constant K_a is altered in the AC4C exciter model (see appendix A.3.1.2). This basic, but illustrative example is chosen to show the general effect of wrong parametrisation on DSA studies. Other incorrect parameters may have less impact on the stability assessment results but alter its trustworthiness and accuracy in general. A sensitivity analysis of model parameters could reveal this impact and is therefore recommended.

According to [200] several indexes are merged to provide a DSA result (see subsection 2.3.3). One relevant index which is merged into the composed DSA result is the angle index (AI). It describes the distance of the synchronous generator rotor angle from an assumed protection setting. Above this limit, the protection relay may separate the machine from the power system. In [200], a value of 120 degree is given as a typical setting by referring to [5]. The AI as defined in [200] is given in equation (5.13). Here, NG is the number of operating generators in the system, $\delta_{i,\max}$ is the maximum rotor angle deviation of the i-th generator occurring during the simulation time is divided by the maximum admissible rotor angle $\delta_{\text{admissible}}$.

$$\mathrm{AI} = \min\left\{1, \max_{i=1\ldots\mathrm{NG}}\left(\frac{\delta_{i,\max}}{\delta_{\text{admissible}}}\right)\right\} \tag{5.13}$$

An exemplary case study for an evaluation of the AI is illustrated in Fig. 5.21. Here, a variation of the exciter gain K_a at power plant GNB is investigated in case that the non-redundant transmission line connecting bus NA and NB is unintentionally tripping at simulation time $t_{\text{sim}} = 100$ ms. The resulting AI varies between 0.02 for $K_a = 20$ to 1.18 for $K_a = 2$. An AI ≥ 1 here represents rotor angle deviation $\delta_{i,\max} \geq 120$ degree. This exemplary case study shows that an accurate statement on the AI and the related dynamic security is only possible in the case of a correctly parameterised exciter model.

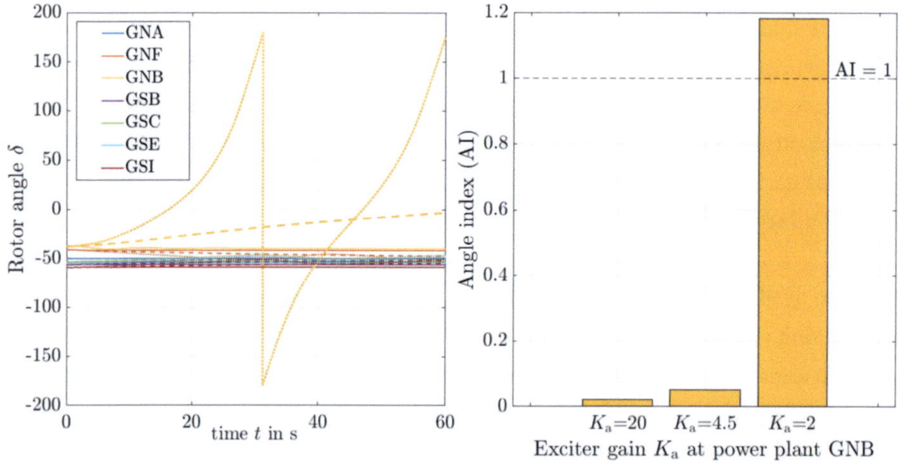

Fig. 5.21: Machine rotor angles in comparison for variant settings of the exciter gain K_a at power plant GNB (left figure): $K_a = 20$ (solid line), $K_a = 4.5$ (dashed line), $K_a = 2$ (dotted line), and derived angle index in comparison (right figure)

Conclusion

The state transition of dynamic power system models highly depends on correct parametrisation of the model structure and the underlying parameter set. The conducted case study shows that an accurate statement on dynamic stability is only possible in the case of a correctly parameterised and validated model. As it is hard to find incorrect parameters manually in large power system models, an automatic validation of these is required. Here it becomes evident that especially the real-time power system operation benefits from the proposed DT-centric EMS architecture (see also subsection 2.2.1). The inherent features enable to react on unpredictable events considering system dynamics during online operation. By increasing confidence in the model results, confidence in the decisions derived from the model and implemented in the power system increases accordingly. Thus, when an immediate response is required, it enables validation of actions before appropriate measures are implemented manually or in an automated scheme. In addition to the applicability for online system operation, all plannable events can also be considered during the phase of operational planning applying the same methods (without the very restrictive computation time constraints of online operation).

5.4.4 Sandbox Assisted System Operation

A validated power system model promises to implement novel system operation methodologies and automation schemes. A numerical case study is chosen to illustrate the sandbox approach. It comprises a load profile variation and the loss of transmission line (TL) NA-NB at simulation time $t_{\mathrm{sim}} = 0.1$ s. This event is considered for all scenarios. Following the loss of TL NA-NB, several remedial actions are tested and validated according to their effect to mitigate the contingency. The effectiveness is illustrated by "traffic light" colour indicators (see also subsection 2.2.2), whereas green represents no violation of boundary limits, orange represents a temporary but acceptable violation of boundary limits, and red represents a severe violation of boundary limits up to system instability. An example plot illustrating the effectiveness of a selected measure is given in Fig. 5.23. The results of the sandbox case study are summarised in Tab. 5.6 and consider the criteria given below:

1. Terminal Voltage limit violation: Voltage magnitude $u_{\mathrm{t}} < 0.9$ or $u_t > 1.1$ p.u.
2. Frequency limit violation: Electrical frequency $f < 49.0$ Hz or $f > 51.0$ Hz
3. Line loading limit violation: TL loading $I_{N\text{-}0} > 0.5 \cdot I_{\mathrm{r}}$; $I_{N\text{-}1} > I_{\mathrm{r}}$ ($I_{\mathrm{r}} = 2.7$ kA)
4. Synchronous machine out of step: Rotor angle deviation exceeds 180 degrees

The selected case study investigates the loss of transmission line NA-NB close to the power plant GNA. The event partially leads to critical states in the benchmark system and to dynamic stability problems as summarised in Tab. 5.6. The numbers in the aggregated result matrix represent the violated criteria according to the enumeration given beforehand. In addition, the colour code corresponds to the severeness of the boundary limit violation. The selected contingency can be positively influenced by applying appropriate curative measures. In some investigated cases, the measure can also negatively influence permissible system states. As the HVDC-System cannot positively influence this particular contingency for topological reasons, the results are not considered in the in the chosen case study.

Due to the characteristics of the benchmark model, the selected case study can only represent the principle of the advantages a DT-centric EMS for arbitrary cases. However, it becomes apparent that an online review of remedial actions is beneficial for system operation to assess both positive and negative effects of certain actions. The principle can be extended to interlock certain actions, to prevent negative impact on system security. The reaction times required to solve the investigated contingency within the given scenario are below human responsiveness. Here, the sandbox assisted system operation illustrates the potential of the proposed DT-centric EMS for automated control. Transferring the online tracked system state into an appropriate validated model in the EMS, allows to investigate the region around current point of operation and to identify possible contingencies, insecure regions of operation stability limits.

A selected case study is illustrated in Fig. 5.22 and Fig. 5.23. The plots illustrate time-domain simulation results for study case F given in Tab. 5.6. The solid lines represent a remedial action applying a fast control action. By switching the capacitor at bus NH at time $t_{sim} = 0.5$ s, the system integrity of the benchmark model can be maintained in comparison to the base case (dashed line), where no action is taken.

Conclusion

Although the remedial action as well as the benchmark model are arbitrary, the case study illustrates the capability of the sandbox functionality the DT-centric EMS provides. To assess all feasible power system states in advance is almost impossible due to the number of possible scenarios and combinations. Furthermore, unforeseeable situations and events cannot be considered during the power system operational planning procedure. While a selected remedial action may be efficient for a particular network configuration, it may not be effective for another configuration. Due to the centralised approach of the proposed DT-centric EMS architecture, the runtime requirements as discussed in subsection 4.2 (summarised in Tab. 4.2) must be met, when the control action is coordinated from the EMS. In most cases, the identification of suitable measures is finished before a critical contingency occurs. Thus, the concept also supports decentral and autonomous approaches. An appropriate logic is then required at the location to control the device independently.

Tab. 5.6: Aggregated results of the sandbox study case (loss of TL NA-NB)

Remedial action following loss of TL NA-NB		t_{sim} in *s*	System load scale factor related to the base case (see appendix A.5)									
			1.0	**0.9**	**0.8**	**0.7**	**0.6**	**0.5**	**0.4**	**0.3**	**0.2**	**0.1**
-	Loss of TL NA-NB (no action)	0.1	1 2 3 4	1 2 3 4	1	1	-	-	-	-	-	-
A	Reconnect TL NA-NB	0.6	1 2 3	1 2 3	2 3	2 3	2	2	-	-	-	-
B	Reconnect TL NA-NB	1.0	1 2 3	1 2 3	1 2 3	1 2 3	2 3	2	2	2	2	3
C	Reconnect TL NA-NB	1.4	1 2 3	1 2 3	1 2 3	1 2 3	2	2	2	-	-	-
D	Reconnect TL NA-NB	1.5	1 2 3 4	1 2 3	1 2 3	1 2 3	2	2	3	-	-	-
E	Set capacitor at bus NH to 250 MVAr	0.5	1 2 3 4	1	-	-	-	-	-	-	-	-
F	Set capacitor at bus NH to 500 MVAr	0.5	1 3	1	1	1	1	1	1	1	1	1
G	Shedding of load at bus NH	0.5	1 2 3 4	1 2 3 4	1	1	-	-	-	-	-	-
H	Shedding of load at bus NH and partial shedding at bus SA	0.5	1 2 3	1 2	1 2	1 2	2	2	-	-	-	-
I	Shedding of load SA and fast curtailment of generation at bus NH	0.5	1 2	1 2	2	-	-	-	-	-	-	-

It can be concluded that the validated DDM instance can serve as a sandbox for power system operation. It allows to steer the power system on a pathway secured by simulations. As a result, power system operation, secured by simulations becomes possible, and DT-assisted partially or even fully automated power system operating routines are enabled.

Due to its high simulation accuracy, of the validated and verified high-fidelity model a DTs provides, a further field of application is the training of machine learning based applications. By providing a unique signature of specific events derived from the validated high-fidelity model, pattern recognition-based decentralised approaches can be implemented as proposed e.g., in [91, 96]. Especially when measurement data is scarce, dedicated simulations of the supervised system can provide synthetic data to train machine learning model approximations. This is a promising field for real-time applications where computational time is scarce and a low fidelity model approximation is sufficient (see also [282]).

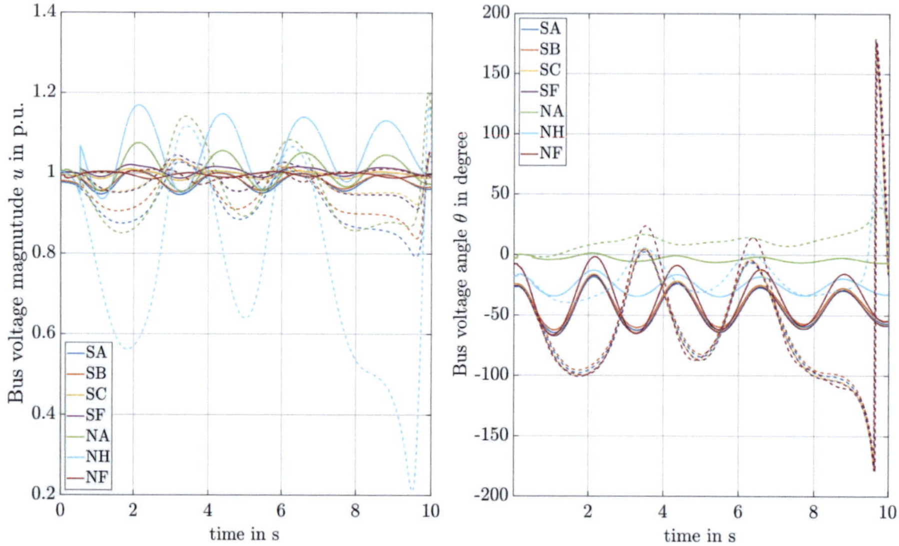

Fig. 5.22: Voltage magnitude u and voltage angle θ resulting for case F (solid line) in comparison to the base case without remedial action (dashed line)

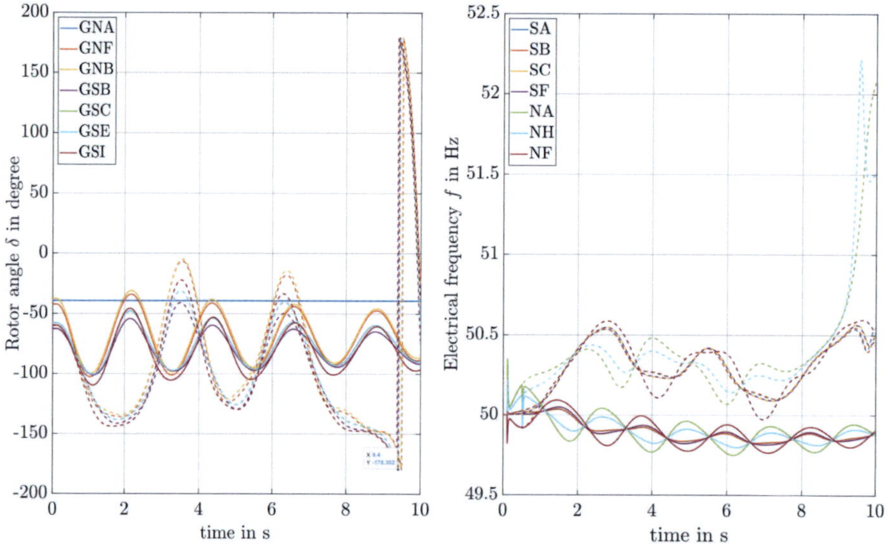

Fig. 5.23: Rotor angle δ and electrical frequency f resulting for case F (solid line) in comparison to the base case without remedial action (dashed line)

6 Conclusion and Outlook

To conclude this thesis, the main findings and research results are summarised within this chapter. Followed by a short assignment of the research subject to the field of practical application, the research questions as formulated in subsection 1.4 are recaptured and answered briefly. Finally, fields for future research activities are identified, and a roadmap towards implementation is outlined.

6.1 Thesis Summary

The power system is in the transition towards a volatile dynamic system, integrating a large amount of renewable energy resources. This implies a higher utilisation of existing power system equipment, and the constant need to adapt to new situations in system operation. The foreseeable necessary functions of future EMS go beyond what can be managed with today's methods, both in terms of data provision and the response time that a human being is capable of. Therefore, a higher degree of power system automation is needed, since the time to react to changes decreases towards timeframes beyond human responsiveness.

To tackle these major challenges, a range of new strategies and novel concepts for future power system operation need to be developed. These include a higher degree of automation, novel system operation strategies that allow automatic adaptation to operational situations, e.g., preventive, or curative system operation approaches, as well as special protection schemes, and continuous control functions. This includes the consideration of the planning time frame in the EMS, and the assessment of dynamic security during online system operation. The novel requirements reveal the key challenge: to set up an appropriate, consistent, and maintainable data model.

A promising approach is the combination of traditional power system models and sensor data. By linking sensor data and models, a continuously adaptive model can be created. This high-fidelity real-time image of the power system can provide new information for system operation. This concept, referred to as a Digital Twin can close an existing applicational gap of EMS in power system control centres. It has been identified as a key concept to reach a higher level of power system automation. Based on these fundamental findings, an advanced architecture for the next generation of energy management systems for monitoring and control of future electric power systems has been designed. The prevailed technologies have been analysed and the requirements to create a DT-centric EMS have been outlined. The feasibility of the proposed approach is demonstrated through a proof of concept and illustrative case studies.

Followed by the introduction and motivation the research work in section 1, an extensive state-of-the-art analysis has been given in section 2. It has been shown that the technological prerequisites already exist to develop and implement the proposed modular and DT-centric EMS architecture. Section 3 outlines the basic functions, and the main inherent characteristics of a DT. A definition of the DT that serves the context of this work is given in subsection 3.1.2. The methods and techniques applied within this research work have been discussed and the requirements to create a DT of the power system are summarised. For the purpose to provide services to power system operation and operational planning, a DT framework is introduced in section 4. It comprises the core components of a DT, including a suitable modelling engine, which allows to mirror the actual power system state to support decision making and to execute simulations to predict future states. To assess electro-mechanical interactions of the power system online, the inherent dynamic models require validation during system operation. Thus, an essential component of the proposed DT framework digital twin is the procedure for model validation and parameter estimation during operation. Therefore, a methodology based on a moving horizon approach has been described (see subsection 3.4.2). It can estimate the parameters of nonlinear equipment models by utilising PMU data streams. Finally, a reference architecture for implementation of the DT-centric EMS is illustrated in subsection 4.4. A concept for validation of the test environment described in subsection 4.3 is given in subsection 4.5.

The hypothesis that a DT and a dedicated modelling engine could become a central element, which provides a new data basis for network analysis applications in the control room has been investigated and validated. Furthermore, a metric to measure the model accuracy is proposed. An important feature of the concept is the moving horizon estimation-based model parameter identification function. It enables to achieve the DTs objective function to create a high-fidelity representation that accurately reflects the dynamic power system states. The key components of the proposed concept are illustrated in Fig. 6.1. In combination with a number crunching real-time simulation platform, the concept allows to create powerful decision support functions, and automated assistant routines, which in turn allow a higher level of power system automation. The feasibility of the proposed concept is examined by proof of concept in section 5.

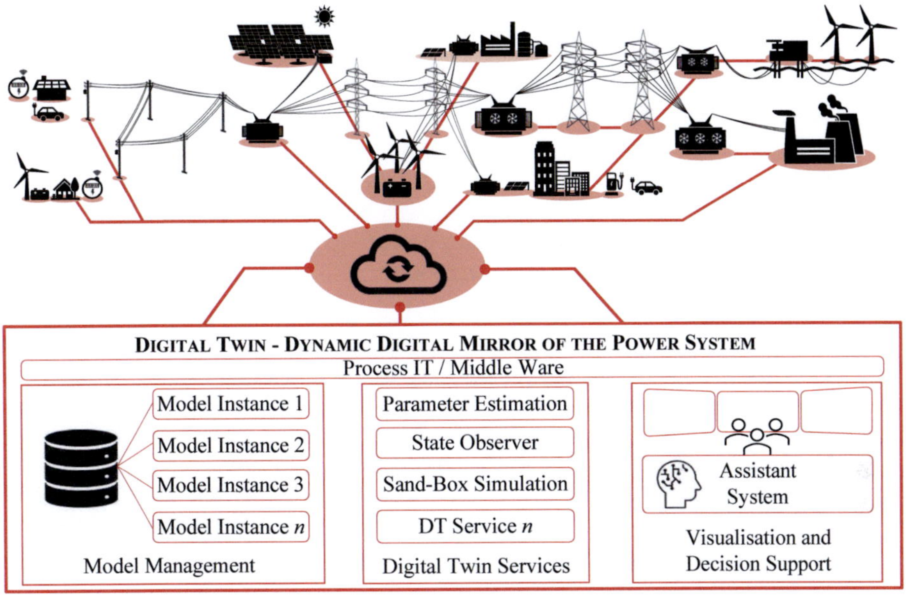

Fig. 6.1: The key components of the proposed concept (upper part inspired by [6])

6.2 Answers to the Research Questions

Several research questions have been formulated at the beginning of this thesis (see subsection 1.4). By answering them briefly in this subsection the main research outcomes of this thesis are summarised.

RQ 1. Which functionalities are required to create a future-proof EMS and how can they be achieved?

The power system is in transition towards a dynamic system which is utilised closer to its physical asset boundaries. This implies the need to adapt to new situations continuously and imposes the requirement of a higher degree of automation to maintain overall system security. The state-of-the-art analysis given in section 2 unveils, that modern control centre EMS inherit complex hardware and software interactions over dedicated ICT with strong security, reliability, and accuracy requirements. These upcoming challenges set the requirements for a future-proof EMS. The proposed DT-centric EMS addresses the identified key requirements by providing the following features and functionalities:

- **An appropriate, consistent, and maintainable data model:**
  The standardised CIM/CGMES based data model supports different use cases, organisational changes, flexible model exchange, as well as the integration of novel applications and methodologies in power system planning and operation. The proposed process IT solution based on the SSOT concept, in conjunction with the developed modelling engine enable iterative growth, including the support of legacy

system components and novel applications. An additional interface to the operational planning and the asset management department enables consistent model maintenance throughout the enterprise or system operator.

- **Enhanced observability and situation awareness:**
  In addition to available observability provided by the SCADA system and WAMS the approach allows a deeper integration of the power system model into the EMS. Available real-time sensor data streams can be utilised to generate a trusted and validated model instance that provides further insights into the power system state including equipment models and controllers.
- **Quicker diagnosis and decision making:**
  A framework has been proposed, that enables to develop powerful decision support tools and assistant systems. Quicker diagnosis and decision making by operators can be achieved by automated decision support functions to avoid reaching the so-called human cognitive barrier during real-time operation. Reaction times below the human responsiveness can be reached by dedicated automation schemes, which utilise the sandbox environment created within the framework for validation of measures before implementation in the real power system in the physical world.
- **Increased trust by automated model validation and recalibration:**
  The proposed framework allows to conduct predictive simulations with higher accuracy. Therefore, an automated model validation and calibration functionality has been proposed, that rises the accuracy of the power system time-domain model. This raises the confidence in the generated decision options, pointing towards the envisioned aim of highly automated power system operation.
- **A modular adaptable EMS architecture:**
  The proposed EMS architecture comprising standardised interfaces to the process and the application modules can serve as a multi-vendor application development platform. The possibility to implement or retrofit assistance systems is ensured. The modular approach suits actual and future business needs and supports varying regulatory frameworks. It simplifies the orchestration and integration of applications and methodologies for different use cases, including support of legacy system components and applications. The additional inherent requirements including the necessary data and information processing have been addressed by parallelisation and middleware approaches. Furthermore, modular approach embraces organisational changes and successive growth. It reduces excessive implementation times and efforts caused by limited developer and power system expert capacities.

RQ 2. What is a Digital Twin, and which characteristics distinguish it from conventional models?

As outlined in subsection 3.1, a *Digital Twin* is a continuously adaptive software-based abstraction of a physical system or object, which is connected in a bidirectional way to the physical system by data streams. As the term is constantly evolving, a definition based on characteristics derived from a literature review has been proposed in subsection 3.1.2 to support the context of this research work. As described in subsection 3.1, the main difference between a conventional digital model and a digital twin is the ability to exchange data bidirectional with the physical object automatically. While a conventional model is updated manually, a DT can alter its underlying model automatically by adaption, utilising the states captured by sensors. An interface to the observed process is a prerequisite to create a DT of the power system. It enables to update states and to tune the model parameters in real time.

An ideal DT represents the case of a high-fidelity model without uncertainties, which means that in practical application the estimated state converges towards the true state of the observed unit, i.e., the estimation error is close to zero along the entire trajectory, provided that the system model is correct. By utilising a DT of a power system, state values can be derived, which are not directly observed by measurements. These can be reported back to the real world to support operational decisions. This fundamental property of a DT appears to be a similarity to the concept of an observer, which is often applied in control engineering. Therefore, it can be concluded that the digital twin represents a special form of an observer, which provides data that can be used purposefully e.g., for future power system operation. An aspect which distinguishes the proposed DT based concept from the traditional observer approach is that it provides a higher flexibility in terms of the model basis. While the observer does only reproduce specific states of a system for a dedicated control purpose, a DT can also reproduce the behaviour of the system in a standalone simulation under various conditions. Its purpose is to mirror the real system as accurate as possible, while providing all necessary information required to maintain secure power system operation.

Today's tools to secure operative security are mainly deterministic and do avoid probabilistic approaches. Therefore, DTs for application in power system operation are based on analytic models. Thus, the model is described by a set of equations, derived from physical laws that can reflect the behaviour of the assets of an electrical power system accurately. Nevertheless, in future powerful assistant functions can be developed based on machine learning methods to exploit the huge amount of data which is recorded during daily operation. DTs based on analytic models can help to create such systems by enriching the training data, especially for uncommon operational scenarios.

RQ 3. What are the key requirements to create a power system DT for real-time applications?

The required interacting components to create a power system DT for real-time applications, have been described in section 4. A DT framework has been proposed, which consists of a data and information model management system connected to the DT database, and a dedicated simulation engine. The process IT based on reliable ICT, connects all components, and provides the relevant data to all applications, services, and functions within the DT framework. The key requirements of the components within the framework have been analysed and are briefly summarised in Tab. 6.1.

Tab. 6.1: Requirements summary to create the proposed DT-centric EMS framework

Component	Identified requirement
ICT and process IT	Compliance with the security requirements for critical infrastructures
	Compliance with the reliability and availability demands for reliable control of the operating resources, i.e., high availability and bandwidth to operate system-critical processes
	Appropriate bandwidth to keep latency low and avoid data loss
	Provide interfaces to the process and application modules applying standardised protocols for power system automation
	Maintainability to manage EMS internal communication and to add or extend components to the DT framework (e.g., by a transaction manager based on publisher – subscriber approach)
	Standardised interfaces to support multivendor application modules
Data and information model management	Maintainable, extendible, and structured data model
	Support of component models of different granularity and model depth
	Standardised model description to allow model exchange
	Model management to maintain a consolidated master model
	Adequate database that supports the processing of real-time data streams
	Capability to connect different data sources to maintain consolidated models
	Secured interfaces to provide information to other departments in the enterprise
Simulation engine	Support of steady state and time-domain simulation
	Flexible power system model interpreter for steady state and time-domain simulation models
	Interface to model management that supports model consistency checks and the adaption of model parameters by model tuning and validation routine
	Capability for (faster than) real-time simulation, parallelised computation of scenarios, and co-simulation allowing data exchange between model instances during runtime

RQ 4. How could the architecture of a Digital Twin-centric control centre EMS look like?

Considering the specific requirements for a future-proof EMS discussed throughout the thesis, a possible architecture for a DT-centric EMS has been proposed in section 4. It comprises a dynamic mirror model, a model tuning function and model validation routine, the DT database and model management platform. These components constitute the core instance of the DT framework and build the foundation for novel automation routines for operation of the electric power system. To distribute the required data and information between all interacting EMS components, a transaction manager is applied. It publishes the necessary data models and information objects to all subscribing modules inside the EMS.

It furthermore provides flexible and maintainable interfaces to all components involved in power system operation. The proposed EMS architecture points towards a consistent data modelling by universal access to the available data. This implies that all related data sources within the enterprise (e.g., in the planning department) are available to the model management system and the functional application modules. The innovations that distinguish the proposed modular DT-centric EMS from a conventionally available EMS can be summarised as follows (see also subsection 4.4):

- A central DT-database, that allows universal access and consistent data modelling, throughout the enterprise,
- A flexible and maintainable process IT, to manage data transfer between all interacting components,
- A modelling engine, which serves as sandbox instance to generate and prove the control options before implementation and enables a higher degree of automation,
- A parameter and state estimation function to create a continuously adaptive power system model, namely the dynamic digital mirror (DDM),
- SCADA and WAMS based information are integrated as a data source for model validation,
- The EMS provides enhanced observability considering system dynamics to maintain dynamic security boundaries and system stability,
- The EMS is enabled for retrofitting of advanced assistant functions that allow a higher degree of automation and can guide operators though system operation.

The core components of proposed architecture described in subsection 4.4 are illustrated in Fig. 6.2.

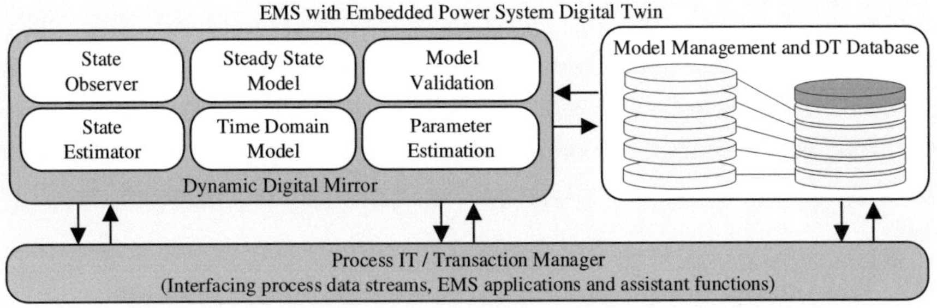

Fig. 6.2: Illustration of the core components of the proposed DT-centric EMS architecture (partly extracted from Fig. 4.9)

RQ 5. Which algorithm is suitable for improving the accuracy of the power system model to develop a dynamic Digital Twin?

To calibrate dynamic power system models, a moving-horizon estimation technique has been proposed. As the MHE technique considers only a defined number of past measurement samples, it combines accurate estimation while maintaining online tractability [164]. The MHE approach is capable to deal with highly non-linear estimation problems and considers state constraints. It has been shown by numerical case studies, that the MHE method is applicable to accurately estimate dynamic model parameters by applying time series data.

RQ 6. Can a Digital Twin-centric EMS increase the degree of automation in system operation?

The tasks of an EMS are to collect and process data to provide a fully comprehensive system state of the observed area of responsibility, which can be applied for decision making. Following the definition of a DT given in this thesis, a Digital Twin-centric EMS comprises a continuously adaptive digital model of the power system, that allows automatic and bidirectional data exchange. By mapping available process data streams to the DT model, an accurate virtual representation of the power system is created. It can serve as an additional data source and a sandbox instance for enhanced model-based decision support and automation functions. The sandbox instance allows to test control actions before implementing them into the real system. The power system DTs inherent characteristics enable the capacity to develop autonomy, which in return points towards a higher degree of automation in power system operation.

The investigations to proof the proposed concept show, that a higher degree of automation of system operation is possible, if the preliminary requirements are fulfilled. The deployment of the DT concept gains trust in the simulation model results and consequently increases trust in the decision options by the higher-level automation functions. It can be concluded that strategies for automatic adaptation to operational situations can be achieved by applying the DT concept as a core instance within the next generation of EMS. These include preventive or curative system operation approaches or continuous control functions.

RQ 7. What impact does the new concept have on operational processes?

A general trend in power system operation is the utilisation of reserves closer to the technical or physical boundary limits. A validated model can help to gain confidence about the distance to the stability or security limits of the modelled system. While notorious inaccurate models will lead to conservative decisions, a high confidence in the model accuracy leads to a higher situation awareness and better and informed operational decisions (see also [287]). Instead of keeping distance to an uncertain boundary by considering "worst-case" reserves, these become utilisable as an additional degree of operational freedom. The consolidated and validated model furthermore allows to improve coordination among power system operators.

A higher degree of automation of the power system inevitably leads to a shift in the responsibility of human control room personnel. The role of the human operator changes towards an automation system engineer, who sets the target values of system variables, defines condition variables and rule sets, and is involved in starting and maintaining the automation system [25]. In addition, new tasks will include the real-time observation of system dynamics, the evaluation of dynamic security limits and the maintenance of the automation system. At a higher level of automation, automation routines can support the operator under normal operating conditions, up to guided problem solving through assistance functions in challenging situations. Ultimately, there is a clear tendency in the scientific community about the future role of operating personnel, which points in the direction of supervisors of autonomous systems.

The development of highly automated systems will set human operators free from repetitive common and frequent tasks, and duties. However, for most of the time in highly automated systems, the human operator would spend the time idling. This is both an opportunity and a challenge. On the one hand, it bears the risk of the *out-of-the-loop syndrome*, which negatively affects the operator's ability to reason and constitutes a serious threat to the safety of the grid. On the other hand, human operators can focus exclusively on handling abnormal grid states [25]. Implementing the proposed concept, allows to create a training system, that can offer daily exercises to the operators or in situations with low workload. Thereby, the operators can train their skills for specific situations, which results in a quieter control room during stressful operational situations.

6.3 Outlook

The proposed concept in this thesis presents an evolutionary advancement of the current state of the art of EMS architectures and functionalities. The steps towards the next generation of EMS have been described in detail and a demonstrator has been developed by implementing the resulting modular architecture in a test environment. The proof of concept is conducted by numerical phasor-based time-domain simulations. However, recent research works have identified, that phasor modelling approaches in close to zero-inertia power systems tend to underestimate the effects of network dynamics on the overall system response [388]. Thus, they indicate that stability criteria are met, while higher order models indicate stability criteria violations. By nature, simplified phasor models cannot reproduce high-frequency transient responses. Therefore, modelling with higher details is necessary, i.e., models which can represent EMT phenomena. As the time step size requirement of

EMT simulations is smaller in comparison to a phasor simulation, their computational effort is accordingly higher. A solution to this problem, considering the framework introduced within this thesis could be the application of co-simulation approaches, i.e., to calculate the network areas and equipment models to be considered in EMT in parallel, and to exchange relevant connecting variables in real time between both simulations. Step size variable integration routines in combination with time-warp techniques, known from time-series modelling, could also provide a relevant solution. These ideas can be accompanied by the vision to create cooperating DTs [389]. In terms of coordinated power system operation, this can help to organise system operation with neighbouring TSOs or DSOs, while reducing manual intervention into operational processes. This automated exchange of contextual information between DT instances enhances the observability for all parties involved in or neighbouring the observed area. The concept furthermore allows to implement all kinds of data driven approaches which rely on unique features or signal pattern. The high-fidelity model represented by the DT in the DT-centric EMS is a prerequisite for such methods [282]. The rising number of operational decisions in future power system operation makes automation and the application of so called artificial intelligence an unavoidable step [41, 248, 390]. A core requirement is the verified viability of such methods and techniques in power system operation [265, 266]. Machine learning may enable the system to handle normal and alert situations because of their frequent occurrences. A naive assumption would be that by the employment of machine learning techniques in operation and for decision support all possible grid states could be managed. Human intervention will still be necessary where training data is lacking. This is expected especially in emergency and extreme situations, which are usually beyond the scenarios for which machine learning systems have been trained. Thus, human intuition and the expertise of senior control room staff will be still required to fill the gaps in the set of training scenarios known to the algorithms. EMS in the power system control centres will constantly evolve to adapt to the challenges of the energy transition [391]. Latest research points towards physics informed neuronal networks [392], where the DT concept can bring in its advantages, which include validated models and automated model maintenance. A possible future research direction may include the generative design approach to automatically create effective RAS [393], or DT assisted system operation [394, 395]. The anticipated development cycles required to implement the proposed concept into future control room EMS are summarised in Tab. 6.2. Starting from the current state of the art, required development stages towards the DT-centric EMS architecture are given, that have also been discussed within this thesis. Further development of the approach is an interdisciplinary challenge. Ingenuity and in-depth methods from mathematics, computer science, classical engineering, cognitive sciences, and psychology are required.

Tab. 6.2: Anticipatory roadmap to future real-world implementations of the DT-centric EMS (partly aligned with [74])

Devel-opment	Model management	System control	Hardware requirements	Decision support
Level 0 (State of the Art)	Data silos and distributed non-consolidated models	Manual control and distributed independent controllers	Standard computing platforms, few PMUs	Manual monitoring, manual execution of security assessment and reaction to events
Level 1	Consolidated models	Generation of preventive and curative control options	Digitally enabled substations	Alarms and security risks integrated into a consolidated user interface
Level 2	Standardised model management and model exchange	Distributed automatic control	Edge computing in substations	Recommended action displayed for human operator
Level 3	Online model validation (component wise)	Partly automated control with operator in the loop	HPC platforms, high PMU coverage	Recommended action validated by simulation
Level 4	Embedded full system Digital Twin	Highly automated control / Model predictive control with operator in the loop	Real-time simulation	Highly automated decision support and integrated operator training sessions
Level 5	Autonomous and cooperative Digital Twins [389]	(Partly) autonomous power systems operation possible [22]	Faster than real-time simulation	Guided system operation by assistant systems [248]; operators become supervisors of autonomous systems

7 References

[1] T. M. Siebel and C. Rice, *Digital transformation: Survive and thrive in an era of mass extinction.* New York, NY: RosettaBooks, 2019.

[2] Cigré JWG C2/B4.38, "Capabilities and requirements definition for power electronics based technology for secure and efficient system operation and control," CIGRE Technical Brochure 821, 2020.

[3] Erik Ørum et al., "Future system inertia," ENTSO-E. Accessed: Dec. 2 2020. [Online]. Available: https://eepublicdownloads.entsoe.eu/clean-documents/Publications/SOC/Nordic/Nordic_report_Future_System_Inertia.pdf

[4] P. Tielens and D. van Hertem, "The relevance of inertia in power systems," *Renewable and Sustainable Energy Reviews*, vol. 55, pp. 999–1009, 2016, doi: 10.1016/j.rser.2015.11.016.

[5] P. S. Kundur, *Power system stability and control.* Chennai, New York, St. Louis, San Francisco, Auckland, Bogotá, Caracas, Kuala Lumpur, Lisbon, London, Madrid, Mexico City, Milan, Montreal, San Juan, Santiago, Singapore, Sydney, Tokyo, Toronto: Mc Graw Hill, 1994.

[6] A. M. Prostejovsky, "Increased Observability in Electric Distribution Grids: Emerging Applications Promoting the Active Role of Distribution System Operators," PhD Dissertation, Department of Electrical Engineering, Technical University of Denmark, 2017.

[7] G. Zhang, "Functional Requirements of Next Generation Control Center Applications," EPRI,, Palo Alto, Jun. 2011. Accessed: Mar. 18 2020. [Online]. Available: https://www.epri.com/research/products/1020058

[8] F. F. Wu, K. Moslehi, and A. Bose, "Power System Control Centers: Past, Present, and Future," *Proc. IEEE*, vol. 93, no. 11, pp. 1890–1908, 2005, doi: 10.1109/JPROC.2005.857499.

[9] P. Zhang, F. Li, and N. Bhatt, "Next-Generation Monitoring, Analysis, and Control for the Future Smart Control Center," *IEEE Trans. Smart Grid*, vol. 1, no. 2, pp. 186–192, 2010, doi: 10.1109/TSG.2010.2053855.

[10] Z. A. Vale, H. Morais, M. Silva, and C. Ramos, "Towards a future SCADA," in *2009 IEEE Power & Energy Society General Meeting*, Calgary, Canada, 2009.

[11] A. Narayan *et al.*, "Towards Future SCADA Systems for ICT-reliant Energy Systems," in *International ETG-Congress 2019; ETG Symposium*, 2019.

[12] K. Tomsovic, D. E. Bakken, V. Venkatasubramanian, and A. Bose, "Designing the Next Generation of Real-Time Control, Communication, and Computations for Large Power Systems," *Proc. IEEE*, vol. 93, no. 5, pp. 965–979, 2005, doi: 10.1109/JPROC.2005.847249.

[13] G. Murray, M. N. Johnstone, and C. Valli, "The convergence of IT and OT in critical infrastructure," Security Research Institute (SRI), Edith Cowan University, Perth, Australian Information Security Management Conference, 2017.

[14] A. Kummerow, S. Nicolai, C. Brosinsky, D. Westermann, A. Naumann, and M. Richter, "Digital-Twin based Services for advanced Monitoring and Control of future power systems," in *2020 IEEE Power & Energy Society General Meeting (PESGM)*, Montreal, QC, Canada, 2020.

[15] Andreas Hauser, "Integrale Netzzustandsanzeige zur Unterstützung der Betriebsführung elektrischer Energieversorgungssysteme," Wuppertal, Univ., Diss., Bergische Universität Wuppertal, Wuppertal, 2000.

[16] C. Schneiders, "Visualisierung des Systemzustandes und Situationserfassung in großräumigen elektrischen Übertragungsnetzen," Dissertation, Fachbereich Elektrotechnik, Informationstechnik, Medientechnik, Bergische Universität Wuppertal, Wuppertal, 2014.

[17] B. F. Wollenberg and T. Sakaguchi, "Artificial intelligence in power system operations," *Proc. IEEE*, vol. 75, no. 12, pp. 1678–1685, 1987, doi: 10.1109/PROC.1987.13935.

[18] J. Giri, M. Parashar, J. Trehern, and V. Madani, "The Situation Room: Control Center Analytics for Enhanced Situational Awareness," *IEEE Power and Energy Mag.*, vol. 10, no. 5, pp. 24–39, 2012, doi: 10.1109/MPE.2012.2205316.

[19] S. Brahma, R. Kavasseri, H. Cao, N. R. Chaudhuri, T. Alexopoulos, and Y. Cui, "Real-Time Identification of Dynamic Events in Power Systems Using PMU Data, and Potential Applications—Models, Promises, and Challenges," *IEEE Trans. Power Delivery*, vol. 32, no. 1, pp. 294–301, 2017, doi: 10.1109/TPWRD.2016.2590961.

[20] R. J. Crouser, L. Franklin, and K. Cook, "Rethinking Visual Analytics for Streaming Data Applications," *IEEE Internet Comput.*, vol. 21, no. 4, pp. 72–76, 2017, doi: 10.1109/MIC.2017.2911428.

[21] M. R. Endsley and D. G. Jones, *Designing for situation awareness: An approach to user-centered design*. Boca Raton, London, New York: CRC Press Taylor & Francis Group, 2012.

[22] V. Biagini, M. Subasic, A. Oudalov, and J. Kreusel, "The autonomous grid: Automation, intelligence and the future of power systems," *Energy Research & Social Science*, vol. 65, p. 101460, 2020, doi: 10.1016/j.erss.2020.101460.

[23] Christoph Brosinsky, Rainer Krebs, Dirk Westermann, "Embedded Digital Twins in future energy management systems: paving the way for automated grid control," *at - Automatisierungstechnik*, vol. 68, no. 9, pp. 1–15, 2020.

[24] C. Brosinsky, T. Sennewald, R. Krebs, and D. Westermann, "Applicational Concept for a Dynamic Power System Mirror in the Control Room," in *Cigré Symposium Aalborg*, 2019.

[25] A. M. Prostejovsky, C. Brosinsky, K. Heussen, D. Westermann, J. Kreusel, and M. Marinelli, "The future role of human operators in highly automated electric power systems," *Electric Power Systems Research*, vol. 175, 2019, doi: 10.1016/j.epsr.2019.105883.

[26] M. Rezkalla, M. Pertl, and M. Marinelli, "Electric power system inertia: Requirements, challenges and solutions," *Electr Eng*, vol. 100, no. 4, pp. 2677–2693, 2018, doi: 10.1007/s00202-018-0739-z.

[27] Lawrence E. Jones, *Renewable Energy Integration Practical Management of Variability, Uncertainty, and Flexibility in Power Grids,* 2nd ed., 2017.

[28] Pacific Northwest National Laboratory (PNNL), "The Emerging Interdependence of the Electric Power Grid & Information and Communication Technology," 2015. Accessed: Dec. 1 2021. [Online]. Available: https://www.pnnl.gov/main/publications/external/technical_reports/PNNL-24643.pdf

[29] J. J.Q. Yu, A. Y.S. Lam, D. J. Hill, and V. O.K. Li, "A unified framework for wide area measurement system planning," *International Journal of Electrical Power & Energy Systems*, vol. 96, pp. 43–51, 2018, doi: 10.1016/j.ijepes.2017.09.032.

[30] M. Panteli and P. Mancarella, "The Grid: Stronger, Bigger, Smarter?: Presenting a Conceptual Framework of Power System Resilience," *IEEE Power and Energy Mag.*, vol. 13, no. 3, pp. 58–66, 2015, doi: 10.1109/MPE.2015.2397334.

[31] M. Panteli and P. Mancarella, "Modeling and Evaluating the Resilience of Critical Electrical Power Infrastructure to Extreme Weather Events," *IEEE Systems Journal*, vol. 11, no. 3, pp. 1733–1742, 2017, doi: 10.1109/JSYST.2015.2389272.

[32] M. Panteli and D. S. Kirschen, "Situation awareness in power systems: Theory, challenges and applications," *Electric Power Systems Research*, vol. 122, pp. 140–151, 2015, doi: 10.1016/j.epsr.2015.01.008.

[33] H. Haes Alhelou, M. Hamedani-Golshan, T. Njenda, and P. Siano, “A Survey on Power System Blackout and Cascading Events: Research Motivations and Challenges,” *Energies*, vol. 12, no. 4, p. 682, 2019, doi: 10.3390/en12040682.

[34] Cigré WG C2.16, “Challenges in the control center (EMS) due to distributed generation and renewables,” Cigré, Technical Brochure 700, 2017.

[35] L. Bainbridge, “Ironies of automation,” *Automatica*, vol. 19, no. 6, pp. 775–779, 1983, doi: 10.1016/0005-1098(83)90046-8.

[36] M. Panteli, P. A. Crossley, D. S. Kirschen, and D. J. Sobajic, “Assessing the Impact of Insufficient Situation Awareness on Power System Operation,” *IEEE Trans. Power Syst.*, vol. 28, no. 3, pp. 2967–2977, 2013, doi: 10.1109/TPWRS.2013.2240705.

[37] M. J. Haass *et al.*, “Toward an Objective Measure of Automation for the Electric Grid,” *Procedia Manufacturing*, vol. 3, pp. 5285–5292, 2015, doi: 10.1016/j.promfg.2015.07.611.

[38] “Commission Regulation (EU) 2017/1485 of 2 August 2017 establishing a guideline on electricity transmission system operation: OJ L 220,” in *Official Journal of the European Union*, 2017. [Online]. Available: http://data.europa.eu/eli/reg/2017/1485/oj

[39] Expert Panel on the separation of the Continental Europe Synchronous Area of 08 January 2021, *Continental Europe Synchronous Area Separation on 08 January 2021: Main Report*. ICS Investigation Expert Panel. [Online]. Available: https://eepublicdownloads.azureedge.net/clean-documents/SOC%20documents/SOC%20Reports/entso-e_CESysSep_Final_Report_210715.pdf (accessed: Jul. 21 2021).

[40] T. E. Dy-Liacco, “Enhancing power system security control,” *IEEE Comput. Appl. Power*, vol. 10, no. 3, pp. 38–41, 1997, doi: 10.1109/67.595291.

[41] C. Brosinsky, T. Sennewald, R. Krebs, D. Westermann, “An Operator Assistant System for Fast and Reliable Decision Support based on a Dynamic Digital Mirror,” in *Cigré Session Paris 2020*.

[42] T. E. Dy Liacco, “Real-time computer control of power systems,” *Proc. IEEE*, vol. 62, no. 7, pp. 884–891, 1974, doi: 10.1109/PROC.1974.9541.

[43] S. K. Khaitan and A. Gupta, *High Performance Computing in Power and Energy Systems*. Berlin, Heidelberg: Springer Berlin Heidelberg, 2013.

[44] Lei Tang, “Dynamic security assessment processing system,” Ph.D. Dissertation, Dep. of Electrical and Computer Engineering, Iowa State University, 2014.

[45] T. H. Demiray, “Simulation of Power System Dynamics using Dynamic Phasor Models,” Ph.D. Dissertation, Swiss Federal Institute of Technology, ETH Zürich, Zürich, Swiss, 2008.

[46] M. Mirz, S. Vogel, G. Reinke, and A. Monti, “DPsim—A dynamic phasor real-time simulator for power systems,” *SoftwareX*, vol. 10, p. 100253, 2019, doi: 10.1016/j.softx.2019.100253.

[47] M. Braun, I. Loeser, H. Hoppe-Oehl, J. König, A. Kubis, M. Lindner, C. Rehtanz, R. Schwerdfeger, “Systematisierung der Autonomiestufen in der Netzbetriebsführung: Systematization of autonomy levels in power system operation,” in *ETG-Fachberichte, ETG-Kongress: Das Gesamtsystem im Fokus der Energiewende 18. - 19. Mai 2021, Online-Veranstaltung,* 1st ed., Berlin: VDE Verlag, 2021.

[48] M. Naglic, “On power system automation: Synchronised measurement technology supported power system situational awareness,” PhD Dissertation, Delft University of Technology, Delft, 2020.

[49] J.-Y. Astic *et al.*, “Control Center Designs: New Functions and Challenges for the Transmission System Operator,” *IEEE Power and Energy Mag.*, vol. 16, no. 2, pp. 57–66, 2018, doi: 10.1109/MPE.2017.2779553.

[50] Cigré Joint Working Group D2/C2.41, "Advanced utility data management and analytics for improved operation situational awareness of EPU operations," Cigré Technical Brochure 732, 2018.

[51] B. Requardt, "Architekturen und Verfahren für modulare Pilotsysteme und Erweiterungen von Netzleitstellen," Dissertation, Kassel, 2021. [Online]. Available: doi://10.17170/kobra-202104073628

[52] E. A. Lee, "Cyber Physical Systems: Design Challenges," in *11th IEEE International Symposium on Object Oriented Real-Time Distributed Computing (ISORC), 2008*, Orlando, Florida, USA, 2008, pp. 363–369.

[53] ARUP, *Digital twin: Towards a meaningful framework.* [Online]. Available: www.arup.com/digitaltwinreport (accessed: Dec. 18 2020).

[54] F. Tao, H. Zhang, A. Liu, and A. Y. C. Nee, "Digital Twin in Industry: State-of-the-Art," *IEEE Trans. Ind. Inf.*, vol. 15, no. 4, pp. 2405–2415, 2019, doi: 10.1109/TII.2018.2873186.

[55] C. Cimino, E. Negri, and L. Fumagalli, "Review of digital twin applications in manufacturing," *Computers in Industry*, vol. 113, p. 103130, 2019, doi: 10.1016/j.compind.2019.103130.

[56] A. Rasheed, O. San, and T. Kvamsdal, "Digital Twin: Values, Challenges and Enablers From a Modeling Perspective," *IEEE Access*, vol. 8, pp. 21980–22012, 2020, doi: 10.1109/ACCESS.2020.2970143.

[57] F. Tao, Q. Qi, L. Wang, and A.Y.C. Nee, "Digital Twins and Cyber–Physical Systems toward Smart Manufacturing and Industry 4.0: Correlation and Comparison," *Engineering*, vol. 5, no. 4, pp. 653–661, 2019, doi: 10.1016/j.eng.2019.01.014.

[58] C. Brosinsky, X. Song, and D. Westermann, "Digital Twin – Concept of a Continuously Adaptive Power System Mirror," in *Internationaler ETG-Kongress, Esslingen am Neckar*, 2019.

[59] P. W.T. Pong *et al.*, "Cyber-enabled grids: Shaping future energy systems," *Advances in Applied Energy*, vol. 1, no. 6, p. 100003, 2021, doi: 10.1016/j.adapen.2020.100003.

[60] L. Wright and S. Davidson, "How to tell the difference between a model and a digital twin," *Adv. Model. and Simul. in Eng. Sci.*, vol. 7, no. 1, A6, 2020, doi: 10.1186/s40323-020-00147-4.

[61] W. Kritzinger, M. Karner, G. Traar, J. Henjes, and W. Sihn, "Digital Twin in manufacturing: A categorical literature review and classification," *IFAC-PapersOnLine*, vol. 51, no. 11, pp. 1016–1022, 2018, doi: 10.1016/j.ifacol.2018.08.474.

[62] J. Trauer, S. Schweigert-Recksiek, C. Engel, K. Spreitzer, and M. Zimmermann, "What is a digital twin: Definitions and insights from an industrial case study in technical product development," *Proc. Des. Soc.: Des. Conf.*, vol. 1, pp. 757–766, 2020, doi: 10.1017/dsd.2020.15.

[63] D. Jones, C. Snider, A. Nassehi, J. Yon, and B. Hicks, "Characterising the Digital Twin: A systematic literature review," *CIRP Journal of Manufacturing Science and Technology*, 2020, doi: 10.1016/j.cirpj.2020.02.002.

[64] B. Hicks, *Industry 4.0 and Digital Twins: Key lessons from NASA.* [Online]. Available: https://www.thefuturefactory.com/blog/24 (accessed: Aug. 29 2021).

[65] *Information technology — Open Systems Interconnection — Basic Reference Model: The Basic Model*, ISO/IEC 7498-1:1994, 1994.

[66] CEN-CENELEC-ETSI Smart Grid Coordination Group, "Smart Grid Reference Architecture," Nov. 2012. Accessed: Feb. 24 2021. [Online]. Available: https://ec.europa.eu/energy/sites/ener/files/documents/xpert_group1_reference_architecture.pdf

[67] *Reference Architecture Model Industrie 4.0 (RAMI4.0)*, DIN SPEC 91345, 2016.

[68] Golftheman, *File:Operating system placement.svg.* [Online]. Available: https://commons.wikimedia.org/wiki/File:Operating_system_placement.svg (accessed: Jun. 30 2021).

[69] C. Brosinsky, D. Westermann, and R. Krebs, "Recent and prospective developments in power system control centers: Adapting the digital twin technology for application in power system control centers," in *2018 IEEE International Energy Conference (ENERGYCON)*, Limassol, Cyprus, 2018, pp. 1–6.

[70] M. R. Endsley and E. S. Connors, "Situation awareness: State of the art," in *2008 IEEE Power and Energy Society General Meeting - Conversion and Delivery of Electrical Energy in the 21st Century*, Pittsburgh, PA, USA, 2008, pp. 1–4.

[71] ENTSO-E, "Dynamic Security Assessment (DSA)," SPD DSA Task Force, RG-CE System Protection & Dynamics Sub Group, Apr. 2017. [Online]. Available: https://docstore.entsoe.eu/Documents/SOC%20documents/Regional_Groups_Continental_Europe/2017/DSA_REPORT_Public.pdf

[72] C. Brosinsky, A. Kummerow, A. Naumann, A. Krönig, S. Balischewski and D. Westermann, "A new development platform for the next generation of power system control center functionalities for hybrid AC-HVDC transmission systems," in *IEEE PES General Meeting, Chicago, USA*, 2017.

[73] A. Kummerow, C. Monsalve, C. Brosinsky, S. Nicolai, and D. Westermann, "A Novel Framework for Synchrophasor Based Online Recognition and Efficient Post-Mortem Analysis of Disturbances in Power Systems," *Applied Sciences*, vol. 10, no. 15, 2020, doi: 10.3390/app10155209.

[74] TenneT TSO B.V., "Control Room of the Future Programme: R&D Roadmap," Version 1.0, Mar. 2021.

[75] *Iconic project: DEEP DIGIT - a digital twin of our power system.* [Online]. Available: https://www.tue.nl/en/research/institutes/eindhoven-institute-for-renewable-energy-systems/systems-integration/ (accessed: Aug. 27 2021).

[76] *Developing a digital twin for the electricity grid.* [Online]. Available: https://www.tudelft.nl/en/eemcs/current/nodes/stories/developing-a-digital-twin-for-the-electricity-grid/ (accessed: Aug. 27 2021).

[77] Energietechnische Gesellschaft (ETG) im VDE e.V., *Digitaler Zwilling in der Elektrizitäts- und Netzwirtschaft: Call for Experts zur Erstellung einer VDE ETG Studie.* [Online]. Available: https://www.vde.com/de/etg/arbeitsgebiete/v2/digitaler-zwilling (accessed: Jun. 1 2022).

[78] *CE-OH Continental Europe Operation Handbook*, 2004.

[79] ENTSO-E, "Network Code on Operational Security," second edition, Sep. 2013.

[80] "Commission Regulation (EU) 2016/1447 of 26 August 2016 establishing a network code on requirements for grid connection of high voltage direct current systems and direct current-connected power park modules: OJ L 241," in *Official Journal of the European Union*, 2016.

[81] *Gesetz über die Elektrizitäts- und Gasversorgung (Energiewirtschaftsgesetz - EnWG).* Accessed: Dec. 14 2020. [Online]. Available: https://www.gesetze-im-internet.de/enwg_2005/

[82] *Gesetz für den Ausbau erneuerbarer Energien (Erneuerbare-Energien-Gesetz - EEG 2017).* Accessed: Dec. 14 2020. [Online]. Available: http://www.gesetze-im-internet.de/eeg_2014/

[83] T. E. Dy Liacco, "The Adaptive Reliability Control System," *IEEE Trans. on Power Apparatus and Syst.*, PAS-86, no. 5, pp. 517–531, 1967, doi: 10.1109/TPAS.1967.291728.

[84] R. Schwerdfeger and D. Westermann, "Vertikaler Netzbetrieb - ein Ansatz zur Koordination von Netzbetriebsinstanzen verschiedener Netzebenen," Dissertation, Technische Universität Ilmenau, 2016. [Online]. Available: http://d-nb.info/113079220X/34

[85] K. Heussen, "Control Architecture Modeling for Future Power Systems," PhD Dissertation, DTU, Kongens, Lyngby, 2011. Accessed: Mar. 21 2020. [Online]. Available: https://backend.orbit.dtu.dk/ws/portalfiles/portal/51229043/Control_Architecture_Modeling.pdf

[86] IEEE Task Force on Definition and Quantification of Resilience, “Comments on the Definition and Quantification of Resilience,” 2018.

[87] E. Ciapessoni, D. Cirio, A. Pitto, M. Panteli, M. van Harte, C. Mak, “Defining power system resilience,” in *ELECTRA*, pp. 32–34.

[88] M. Panteli, D. N. Trakas, P. Mancarella, and N. D. Hatziargyriou, “Power Systems Resilience Assessment: Hardening and Smart Operational Enhancement Strategies,” *Proc. IEEE*, vol. 105, no. 7, pp. 1202–1213, 2017, doi: 10.1109/JPROC.2017.2691357.

[89] ENTSO-E, Subgroup "Protection and Dynamics", “Report on Special Protection Schemes,” Brussels, Mar. 2012. Accessed: Oct. 24 2019. [Online]. Available: https://www.en-tsoe.eu/fileadmin/user_upload/_library/publications/en-tsoe/RG_SOC_CE/120425_RG_CE_TOP_06.5_D.2_SPS_report_1_.pdf

[90] L. H. Fink and K. Carlsen, “Operating under stress and strain [electrical power systems control under emergency conditions],” *IEEE Spectr.*, vol. 15, no. 3, pp. 48–53, 1978, doi: 10.1109/MSPEC.1978.6369445.

[91] F. Sass, T. Sennewald, and D. Westermann, “Automated Corrective Actions by VSC-HVDC-systems: A novel Remedial Action Scheme,” *IEEE Trans. Power Syst.*, pp. 1–10, 2019, doi: 10.1109/TPWRS.2019.2928887.

[92] North American Electric Reliability Cooperation (NERC), *PRC-012-2 – Remedial Action Schemes (Final draft).* [Online]. Available: https://www.nerc.com/pa/Stand/Prjct201005_3RmdialActnSchmsPhase3ofPrtctnSystmsDL/PRC-012-2_04182016_clean_team_final.pdf (accessed: Oct. 24 2019).

[93] NERC, “Remedial Action Scheme Definition Development: Background and Frequently Asked Questions,” Project 2010-05.2 – Special Protection System, Jun. 2014. Accessed: Jan. 18 2019. [Online]. Available: https://www.nerc.com/pa/Stand/Prjct201005_2SpclPrtctnSstmPhs2/FAQ_RAS_Definition_0604_final.pdf

[94] J. McCalley, O. Oluwaseyi, V. Krishnan, R. Dai, C. Singh, and K. Jiang, “System Protection Schemes: Limitations, Risks, and Management,” 2010.

[95] D. Westermann, S. Schlegel, F. Sass, R. Schwerdfeger, A. Wasserrab, U. Häger, S. Dalhues, C. Biele, A. Kubis, J. Hachenberger, “Curative actions in the power system operation to 2030,” in *Internationaler ETG Kongress 2019, Esslingen*, 2019.

[96] F. Sass, “Beitrag zur Systemsicherheit durch kurative Maßnahmen von HGÜ-Systemen,” Dissertation, Fachgebiet Elektrische Energieversorgung, Technische Universität Ilmenau, Ilmenau, 2019.

[97] ENTSO-E, “Supporting Document for the Network Code on Operational Security,” Brussels, 2013. Accessed: Sep. 21 2020. [Online]. Available: https://eepublicdownloads.azureedge.net/clean-documents/pre2015/resources/OS_NC/130924-AS-NC_OS_Supporting_Document_2nd_Edition_final.pdf

[98] J. H. Obradovich, “Understanding Cognitive and Collaborative Work: Observations in an Electric Transmission Operations Control Center,” *Proceedings of the Human Factors and Ergonomics Society Annual Meeting*, vol. 55, no. 1, pp. 247–251, 2011, doi: 10.1177/1071181311551051.

[99] J. Warichet, “From the Measurement of Synchrophasors to the Identification of Inter--Area Oscillations in Power Transmission Systems,” Dissertation, Faculty of Applied Sciences, Université Libre, Brüssel, 2013.

[100] North American Electric Reliability Corporation (NERC), *System Operator Certification Exam Content.* [Online]. Available: https://www.nerc.com/pa/Train/SysOpCert/Pages/default.aspx (accessed: Feb. 27 2019).

[101] Cigré WG C2.34, "Capabilities and requirements of a control centre in the 21st century - functional and human resources view," ELECTRA No. 289, 2016.

[102] D. Cole and P. Grossman, *The End of a Natural Monopoly: Deregulation and Competition in the Electric Power Industry*: Routledge, 2003.

[103] K. Gugler, M. Rammerstorfer, and S. Schmitt, "Ownership unbundling and investment in electricity markets — A cross country study," *Energy Economics*, vol. 40, pp. 702–713, 2013, doi: 10.1016/j.eneco.2013.08.022.

[104] S. Ruester, S. Schwenen, C. Batlle, and I. Pérez-Arriaga, "From distribution networks to smart distribution systems: Rethinking the regulation of European electricity DSOs," *Utilities Policy*, vol. 31, pp. 229–237, 2014, doi: 10.1016/j.jup.2014.03.007.

[105] Union of the Electricity Industry (Eurelectric), *Eurelectric's vision about the role of Distribution System Operators (DSOs)*. Technical report. [Online]. Available: https://www.eurelectric.org/media/2000/dso_vision_final_100216_web-2016-030-0092-01-e.pdf (accessed: Jul. 12 2019).

[106] Cigré Joint Working Group C2/C6.36, "System operation emphasizing DSO/TSO interaction and coordination," CIGRE Technical Brochure 733, 2018.

[107] IEC TS 61970-600-1, *Energy management system application program interface (EMS-API)*, 1st ed. Geneva, Switzerland: International Electrotechnical Commission, 2017.

[108] *Energy management system application program interface (EMS-API) - Part 600-2: Common Grid Model Exchange Specification (CGMES) - Exchange profiles specification*, IEC TS 61970-600-2, 2017.

[109] F. Jańıček, M. Jedinák, and I. Šulc, "Awareness System Implemented in the European Network," *Journal of Electrical Engineering*, vol. 65, no. 5, pp. 320–324, 2014, doi: 10.2478/jee-2014-0053.

[110] W. H. WELLSSOW, et al., "TSO / DSO Cooperation and Interactions in Systems with Very High Shares of Renewable Generation: C2-207," in *Cigré Session Paris 2020*.

[111] A. Naumann and Z. A. Styczynski, "Leitwarte im Smart Grid," Dissertation, Fak. für Elektrotechnik und Informationstechnik,, Otto von Guericke Univiersität Magdeburg (OvGU), Magdeburg, 2012.

[112] *Power systems management and associated information exchange. Reference architecture*, PD IEC TR 62357-1:2016, 2016.

[113] M. Uslar, M. Specht, S. Rohjans, J. Trefke, and J. M. Vasquez Gonzalez, *The Common Information Model CIM: IEC 61968/61970 and 62325 – A Practical Introduction to the CIM*. Berlin, Heidelberg: Springer Berlin Heidelberg, 2012.

[114] *Energy management system application program interface (EMS-API) - Common Grid Model Exchange Specification (CGMES)*, IEC 61970-CGMES:2022, 2022.

[115] ENTSO-E, *CGMES Roadmap*. [Online]. Available: https://docstore.entsoe.eu/Documents/CIM_documents/CGMES_Roadmap_2018-SOC.pdf

[116] K. Schwarz, "Telecontrol Standard IEC 60870-6 TASE.2 Globally Adopted," in *Fieldbus Technology*, D. Dietrich, H. Schweinzer, and P. Neumann, Eds., Vienna: Springer Vienna, 1999, pp. 38–45.

[117] *Communication networks and systems for power utility automation - Part 1: Introduction and overview*, IEC TR 61850-1:2013, 2013.

[118] *Telecontrol equipment and systems – Part 5-104: Transmission protocols – Network access for IEC 60870-5-101 using standard transport profiles*, IEC 60870-5-104, 2006.

[119] *Telecontrol equipment and systems. Part 5. Transmission protocol. Section 101. Companion standard for basic telecontrol tasks*, IEC 60870-5-101:2006, 2006.

[120] *IEEE Standard for Electric Power Systems Communications-Distributed Network Protocol (DNP3)*, 2012.

[121] K. Schwarz, "Comparison of IEC 60870-5-101/-103/-104 and IEC 60870-6-TASE.2 with IEC 61850," Version 4, Jul. 2012.

[122] *Power systems management and associated information exchange – Data and communications security Part 1: Communication network and system securityIntroduction to security issues*, IEC TS 62351-1, 2007.

[123] K.-P. Brand, V. Lohmann, and W. Wimmer, *Substation automation handbook: Comprehensive description of substation automation and the coordination with network operation to obtain both performance and cost benefits by enabling enhanced power system management*. Bremgarten: Utility Automation Consulting Lohmann, 2003.

[124] R. Hunt, B. Flynn, and T. Smith, "The Substation of the Future: Moving Toward a Digital Solution," *IEEE Power and Energy Mag.*, vol. 17, no. 4, pp. 47–55, 2019, doi: 10.1109/MPE.2019.2908122.

[125] J. Giri, "Proactive Management of the Future Grid," *IEEE Power Energy Technol. Syst. J.*, vol. 2, no. 2, pp. 43–52, 2015, doi: 10.1109/JPETS.2015.2408212.

[126] A. J. Schwab, *Elektroenergiesysteme: Erzeugung, Übertragung und Verteilung elektrischer Energie*, 4th ed. Berlin, Heidelberg: Springer Berlin Heidelberg, 2015. [Online]. Available: http://gbv.eblib.com/patron/FullRecord.aspx?p=4068764

[127] E. Handschin and A. Petroianu, *Energy management systems: Operation and control of electric energy transmission systems*. Berlin: Springer, 1991.

[128] F. Maghsoodlou, R. Masiello, and T. Ray, "Energy management systems," *IEEE Power and Energy Mag.*, vol. 2, no. 5, pp. 49–57, 2004, doi: 10.1109/MPAE.2004.1338122.

[129] D. Rumpel and J. R. Sun, *Netzleittechnik: Informationstechnik für den Betrieb Elektrischer Netze*: Springer, 1989.

[130] C. Brosinsky, F. Sass, T. Sennewald, R. Krebs, and D. Westermann, "HVAC/HVDC Control Center - Test and Demonstrator System," in *International ETG Congress 2017*, 2017, pp. 1–6.

[131] H. Dommel and W. Tinney, "Optimal Power Flow Solutions," *IEEE Trans. on Power Apparatus and Syst.*, PAS-87, no. 10, pp. 1866–1876, 1968, doi: 10.1109/TPAS.1968.292150.

[132] A. Monticelli, M. V. F. Pereira, and S. Granville, "Security-Constrained Optimal Power Flow with Post-Contingency Corrective Rescheduling," *IEEE Trans. Power Syst.*, vol. 2, no. 1, pp. 175–180, 1987, doi: 10.1109/TPWRS.1987.4335095.

[133] *IEEE Standard for SCADA and Automation Systems*, IEEE C37.1, 2007.

[134] K. Heuck, K.-D. Dettmann, and D. Schulz, *Elektrische Energieversorgung: Erzeugung, Übertragung und Verteilung elektrischer Energie für Studium und Praxis*, 9th ed. Wiesbaden: Springer Vieweg, 2013.

[135] E.-G. Tietze, *Netzleittechnik*, 2nd ed. Berlin: VDE-Verl., 2006.

[136] F. Schweppe and J. Wildes, "Power System Static-State Estimation, Part I: Exact Model," *IEEE Trans. on Power Apparatus and Syst.*, PAS-89, no. 1, pp. 120–125, 1970, doi: 10.1109/TPAS.1970.292678.

[137] F. Schweppe and D. Rom, "Power System Static-State Estimation, Part II: Approximate Model," *IEEE Trans. on Power Apparatus and Syst.*, PAS-89, no. 1, pp. 125–130, 1970, doi: 10.1109/TPAS.1970.292679.

[138] F. Schweppe, "Power System Static-State Estimation, Part III: Implementation," *IEEE Trans. on Power Apparatus and Syst.*, PAS-89, no. 1, pp. 130–135, 1970, doi: 10.1109/TPAS.1970.292680.

[139] F. C. Schweppe and E. J. Handschin, "Static state estimation in electric power systems," *Proc. IEEE*, vol. 62, no. 7, pp. 972–982, 1974, doi: 10.1109/PROC.1974.9549.

[140] T. E. Dy Liacco, “State estimation in control centres,” *International Journal of Electrical Power & Energy Systems*, vol. 5, no. 4, pp. 218–221, 1983, doi: 10.1016/0142-0615(83)90022-4.

[141] A. Bose and K. A. Clements, “Real-time modeling of power networks,” *Proc. IEEE*, vol. 75, no. 12, pp. 1607–1622, 1987, doi: 10.1109/PROC.1987.13930.

[142] L. Mili, M. G. Cheniae, N. S. Vichare, and P. J. Rousseeuw, “Robust state estimation based on projection statistics [of power systems],” *IEEE Trans. Power Syst.*, vol. 11, no. 2, pp. 1118–1127, 1996, doi: 10.1109/59.496203.

[143] A. Monticelli, *State estimation in electric power systems: A generalized approach*. Boston: Kluwer Academic Publishers, 1999. [Online]. Available: http://www.loc.gov/catdir/enhancements/fy0821/99020753-d.html

[144] A. Abur and A. G. Exposito, *Power System State Estimation: Theory and Implementation*. Hoboken: Marcel Dekker Inc, 2004.

[145] F. F. Wu, “Power system state estimation: A survey,” *International Journal of Electrical Power & Energy Systems*, vol. 12, no. 2, pp. 80–87, 1990, doi: 10.1016/0142-0615(90)90003-T.

[146] K. Das, J. Hazra, D. P. Seetharam, R. K. Reddi, and A. K. Sinha, “Real-time hybrid state estimation incorporating SCADA and PMU measurements,” in *2012 3rd IEEE PES Innovative Smart Grid Technologies Europe (ISGT Europe)*, Berlin, Germany, 2012, pp. 1–8.

[147] D. Tomašević, S. Avdaković, Z. Bajramović, and I. Džananović, “Comparison of Different Techniques for Power System State Estimation,” in *Lecture Notes in Networks and Systems, Advanced Technologies, Systems, and Applications III: Proceedings of the International Symposium on Innovative and Interdisciplinary Applications of Advanced Technologies (IAT), Volume 1*, S. Avdakovic, Ed., 2019.

[148] Z. JIN, P. Wall, Y. Chen, J. Yu, S. Chakrabarti, and V. Terzija, “Analysis of Hybrid State Estimators: Accuracy and Convergence of Estimator Formulations,” *IEEE Trans. Power Syst.*, vol. 34, no. 4, pp. 2565–2576, 2019, doi: 10.1109/TPWRS.2018.2871192.

[149] M. Gol and A. Abur, “A Hybrid State Estimator For Systems With Limited Number of PMUs,” *IEEE Trans. Power Syst.*, vol. 30, no. 3, pp. 1511–1517, 2015, doi: 10.1109/TPWRS.2014.2344012.

[150] A. S. Dobakhshari, S. Azizi, M. Paolone, and V. Terzija, “Ultra Fast Linear State Estimation Utilizing SCADA Measurements,” *IEEE Trans. Power Syst.*, vol. 34, no. 4, pp. 2622–2631, 2019, doi: 10.1109/TPWRS.2019.2894518.

[151] V. Donde, X. Feng, I. Segerqvist, and M. Callavik, “Distributed State Estimation of Hybrid AC/HVDC Grids by Network Decomposition,” *IEEE Trans. Smart Grid*, vol. 7, no. 2, pp. 974–981, 2016, doi: 10.1109/TSG.2015.2457932.

[152] W. Jiang, V. Vittal, and G. T. Heydt, “A Distributed State Estimator Utilizing Synchronized Phasor Measurements,” *IEEE Trans. Power Syst.*, vol. 22, no. 2, pp. 563–571, 2007, doi: 10.1109/TPWRS.2007.894859.

[153] Y. Sun, M. Fu, and H. Zhang, “Performance comparison of distributed state estimation algorithms for power systems,” *J Syst Sci Complex*, vol. 30, no. 3, pp. 595–615, 2017, doi: 10.1007/s11424-017-6062-3.

[154] B. H. Bringeland, “A Substation Level State Estimator for Local Data Processing: Algorithms for Power System Monitoring,” Master Thesis, Department of Electric Power Engineering, Norwegian University of Science and Technology (NTNU), Trondheim, Norway, 2017.

[155] P. M. DeRusso, *State variables for engineers,* 2nd ed. New York: Wiley, 1998.

[156] J.-N. Juang, *Applied system identification*. Englewood Cliffs, NJ: Prentice Hall, 1994.

[157] G. Ellis, *Control system design guide: Using your computer to understand and diagnose feedback controllers,* 4th ed. Amsterdam: Elsevier, 2012.

[158] R. E. Kalman, "A New Approach to Linear Filtering and Prediction Problems," *Journal of Basic Engineering*, vol. 82, no. 1, pp. 35–45, 1960, doi: 10.1115/1.3662552.

[159] M. S. Grewal and A. P. Andrews, "Applications of Kalman Filtering in Aerospace 1960 to the Present [Historical Perspectives]," *IEEE Control Syst.*, vol. 30, no. 3, pp. 69–78, 2010, doi: 10.1109/MCS.2010.936465.

[160] R. E. Kalman and R. S. Bucy, "New Results in Linear Filtering and Prediction Theory," *Journal of Basic Engineering*, vol. 83, no. 1, pp. 95–108, 1961, doi: 10.1115/1.3658902.

[161] J. D. Hedengren and A. N. Eaton, "Overview of estimation methods for industrial dynamic systems," *Optim Eng*, vol. 18, no. 1, pp. 155–178, 2017, doi: 10.1007/s11081-015-9295-9.

[162] S. J. Julier and J. K. Uhlmann, "Unscented Filtering and Nonlinear Estimation," *Proc. IEEE*, vol. 92, no. 3, pp. 401–422, 2004, doi: 10.1109/JPROC.2003.823141.

[163] E. L. Haseltine and J. B. Rawlings, "Critical Evaluation of Extended Kalman Filtering and Moving-Horizon Estimation," *Ind. Eng. Chem. Res.*, vol. 44, no. 8, pp. 2451–2460, 2005, doi: 10.1021/ie034308l.

[164] J. B. Rawlings, D. Q. Mayne, and M. M. Diehl, *Model predictive control: Theory, computation, and design,* 2nd ed. Madison, Wisconsin: Nob Hill Publishing, 2017.

[165] J. Zhao *et al.,* "Power System Dynamic State Estimation: Motivations, Definitions, Methodologies, and Future Work," *IEEE Trans. Power Syst.*, vol. 34, no. 4, pp. 3188–3198, 2019, doi: 10.1109/TPWRS.2019.2894769.

[166] J. Zhao *et al.,* "Roles of Dynamic State Estimation in Power System Modeling, Monitoring and Operation," *IEEE Trans. Power Syst.*, p. 1, 2020, doi: 10.1109/TPWRS.2020.3028047.

[167] IEEE Power System Operation, Planning and Economics (PSOPE) Committee: Task Force on Power System Dynamic State and Parameter Estimation, "Power System Dynamic State and Parameter Estimation-Transition to Power Electronics-Dominated Clean Energy Systems," Technical Report PES-TR88, 2021.

[168] J. Zhao, M. Netto, and L. Mili, "A Robust Iterated Extended Kalman Filter for Power System Dynamic State Estimation," *IEEE Trans. Power Syst.*, vol. 32, no. 4, pp. 3205–3216, 2017, doi: 10.1109/TPWRS.2016.2628344.

[169] X. Wang, J. Zhao, V. Terzija, and S. Wang, "Fast robust power system dynamic state estimation using model transformation," *International Journal of Electrical Power & Energy Systems*, vol. 114, p. 105390, 2020, doi: 10.1016/j.ijepes.2019.105390.

[170] J. Zhao and L. Mili, "Robust Unscented Kalman Filter for Power System Dynamic State Estimation With Unknown Noise Statistics," *IEEE Trans. Smart Grid*, vol. 10, no. 2, pp. 1215–1224, 2019, doi: 10.1109/TSG.2017.2761452.

[171] T. Wu, C.-Y. Chung, and I. Kamwa, "A Fast State Estimator for Systems Including Limited Number of PMUs," *IEEE Trans. Power Syst.*, 2017, doi: 10.1109/TPWRS.2017.2673857.

[172] L. Jin, R. Kumar, and N. Elia, "Model Predictive Control-Based Real-Time Power System Protection Schemes," *IEEE Trans. Power Syst.*, vol. 25, no. 2, pp. 988–998, 2010, doi: 10.1109/TPWRS.2009.2034748.

[173] A. Fuchs, M. Imhof, T. Demiray, and M. Morari, "Stabilization of Large Power Systems Using VSC–HVDC and Model Predictive Control," *IEEE Trans. Power Delivery*, vol. 29, no. 1, pp. 480–488, 2014, doi: 10.1109/TPWRD.2013.2280467.

[174] K. Sun, J. Qi, and W. Kang, “Power system observability and dynamic state estimation for stability monitoring using synchrophasor measurements,” *Control Engineering Practice*, vol. 53, pp. 160–172, 2016, doi: 10.1016/j.conengprac.2016.01.013.

[175] Y. Xu, Z. Y. Dong, J. H. Zhao, P. Zhang, and K. P. Wong, “A Reliable Intelligent System for Real-Time Dynamic Security Assessment of Power Systems,” *IEEE Trans. Power Syst.*, vol. 27, no. 3, pp. 1253–1263, 2012, doi: 10.1109/TPWRS.2012.2183899.

[176] H. Liu, Z. Hu, and Y. Song, “Lyapunov-Based Decentralized Excitation Control for Global Asymptotic Stability and Voltage Regulation of Multi-Machine Power Systems,” *IEEE Trans. Power Syst.*, vol. 27, no. 4, pp. 2262–2270, 2012, doi: 10.1109/TPWRS.2012.2196716.

[177] R. Yan, Z. Dong, T. K. Saha, and R. Majumder, “A power system nonlinear adaptive decentralized controller design,” *Automatica*, vol. 46, no. 2, pp. 330–336, 2010, doi: 10.1016/j.automatica.2009.10.020.

[178] H. Tebianian and B. Jeyasurya, “Dynamic state estimation in power systems using Kalman filters,” in *2013 IEEE Electrical Power & Energy Conference*, Halifax, NS, Canada, 2013, pp. 1–5.

[179] E. Ghahremani and I. Kamwa, “Dynamic State Estimation in Power System by Applying the Extended Kalman Filter With Unknown Inputs to Phasor Measurements,” *IEEE Trans. Power Syst.*, vol. 26, no. 4, pp. 2556–2566, 2011, doi: 10.1109/TPWRS.2011.2145396.

[180] J. Qi, K. Sun, J. Wang, and H. Liu, “Dynamic State Estimation for Multi-Machine Power System by Unscented Kalman Filter With Enhanced Numerical Stability,” *IEEE Trans. Smart Grid*, vol. 9, no. 2, pp. 1184–1196, 2018, doi: 10.1109/TSG.2016.2580584.

[181] H. Tebianian and B. Jeyasurya, “Dynamic state estimation in power systems: Modeling, and challenges,” *Electric Power Systems Research*, vol. 121, pp. 109–114, 2015, doi: 10.1016/j.epsr.2014.12.005.

[182] G. Valverde and V. Terzija, “Unscented Kalman filter for power system dynamic state estimation,” *IET Generation, Transmission & Distribution*, vol. 5, no. 1, p. 29, 2011, doi: 10.1049/iet-gtd.2010.0210.

[183] A. K. Singh and B. C. Pal, “Decentralized Robust Dynamic State Estimation in Power Systems Using Instrument Transformers,” *IEEE Trans. Signal Process.*, vol. 66, no. 6, pp. 1541–1550, 2018, doi: 10.1109/TSP.2017.2788424.

[184] E. Ghahremani and I. Kamwa, “Local and Wide-Area PMU-Based Decentralized Dynamic State Estimation in Multi-Machine Power Systems,” *IEEE Trans. Power Syst.*, vol. 31, no. 1, pp. 547–562, 2016, doi: 10.1109/TPWRS.2015.2400633.

[185] D. Simon, *Optimal state estimation: Kalman, H [infinity], and nonlinear approaches*. Hoboken, NJ: Wiley-Interscience, 2006.

[186] A. K. Singh and B. C. Pal, *Dynamic estimation and control of power systems*. London, United Kingdom: Academic Press an imprint of Elsevier, 2019.

[187] N. Zhou, D. Meng, Z. Huang, and G. Welch, “Dynamic State Estimation of a Synchronous Machine Using PMU Data: A Comparative Study,” *IEEE Trans. Smart Grid*, vol. 6, no. 1, pp. 450–460, 2015, doi: 10.1109/TSG.2014.2345698.

[188] IEEE/CIGRE Joint Task Force on Stability Terms and Definitions, “Definition and Classification of Power System Stability IEEE/CIGRE Joint Task Force on Stability Terms and Definitions,” *IEEE Trans. Power Syst.*, vol. 19, no. 3, pp. 1387–1401, 2004, doi: 10.1109/TPWRS.2004.825981.

[189] N. Hatziargyriou *et al.*, “Definition and Classification of Power System Stability Revisited & Extended,” *IEEE Trans. Power Syst.*, p. 1, 2020, doi: 10.1109/TPWRS.2020.3041774.

[190] Cigré WG C4.601, "Review of on-line dynamic security assessment tools and techniques," Technical Brochure No. 325, 2007.

[191] B. Stott and E. Hobson, "Power System Security Control Calculations Using Linear Programming, Part I," *IEEE Trans. on Power Apparatus and Syst.*, PAS-97, no. 5, pp. 1713–1720, 1978, doi: 10.1109/TPAS.1978.354664.

[192] B. Stott and E. Hobson, "Power System Security Control Calculations Using Linear Programming, Part II," *IEEE Trans. on Power Apparatus and Syst.*, PAS-97, no. 5, pp. 1721–1731, 1978, doi: 10.1109/TPAS.1978.354665.

[193] L. Wu, J. Gao, Y. Wang, and R. G. Harley, "A survey of contingency analysis regarding steady state security of a power system," in *2017 North American Power Symposium (NAPS)*, Morgantown, WV, 2017, pp. 1–6.

[194] D. Semitekos and N. Avouris, "Steady State Contingency analysis of electrical networks using machine learning techniques," in *Artificial Intelligence Applications and Innovations*, 2006, pp. 281–289.

[195] S. C. Savulescu, *Real-Time Stability in Power Systems: Techniques for Early Detection of the Risk of Blackout,* 2nd ed. Cham: Springer International Publishing, 2014.

[196] A. A. Fouad, F. Aboytes, V. F. Carvalho, S. L. Corey, K. J. Dhir, and R. Vierra, "Dynamic security assessment practices in North America," *IEEE Trans. Power Syst.*, vol. 3, no. 3, pp. 1310–1321, 1988, doi: 10.1109/59.14597.

[197] U. Kerin, C. Heyde, R. Krebs, and E. Lerch, "Real-time dynamic security assessment of power grids," *Eur. Phys. J. Spec. Top.*, vol. 223, no. 12, pp. 2503–2516, 2014, doi: 10.1140/epjst/e2014-02272-1.

[198] C. O. Heyde, *Dynamic voltage security assessment for on-line control room application: (Dynamische Spannungsstabilitätsrechnungen als online Entscheidungsgrundlage für die Leitwarte).* Magdeburg, Univ., Fak. für Elektrotechnik und Informationstechnik, Diss., 2010, 1st ed. Magdeburg: Univ, 2010.

[199] Uroš Kerin, "Development of Tools for Dynamic Security Assessment of Electric Power Systems," PhD Dissertation, Faculty of Electrical Engineering, UNIVERSITY OF LJUBLJANA, Ljubljana, Slovenia, 2010.

[200] J. M. Gimenez Alvarez and P. E. Mercado, "Online Inference of the Dynamic Security Level of Power Systems Using Fuzzy Techniques," *IEEE Trans. Power Syst.*, vol. 22, no. 2, pp. 717–726, 2007, doi: 10.1109/TPWRS.2007.895161.

[201] K. Morison, L. Wang, and P. Kundur, "Power system security assessment," *IEEE Power and Energy Mag.*, vol. 2, no. 5, pp. 30–39, 2004, doi: 10.1109/MPAE.2004.1338120.

[202] J. T. G. Weckesser, "On-line Dynamic Security Assessment in Power Systems," Disseration, Department of Electrical Engineering, Technical University of Denmark, 2014.

[203] EPRI, "Implementation of Dynamic Security Assessment at Northern States Power: Technical Report," Palo Alto, 2001.

[204] O. Ruhle and E. Lerch, "Ranking of system contingencies in DSA systems - first experiences," in *IEEE PES General Meeting*, Minneapolis, MN, 2010, pp. 1–6.

[205] M. Shahraeini and M. Hossein, "Wide Area Measurement Systems," in *Advanced Topics in Measurements*, M. Z. Haq, Ed.: InTech, 2012. [Online]. Available: http://www.intechopen.com/books/advanced-topics-in-measurements/wide-area-measurement-systems

[206] C. P. Steinmetz, "Complex Quantities and Their Use in Electrical Engineering," in *Proceedings of the international electrical congress*, 1893, pp. 33–74.

[207] M. Richter, "PMU-basierte Zustandsabschätzung in Smart Distribution," Dissertation, Fakultät für Elektrotechnik und Informationstechnik, Otto-von-Guericke-Universität Magdeburg, 2016.

[208] A. Primadianto and C.-N. Lu, "A Review on Distribution System State Estimation," *IEEE Trans. Power Syst.*, vol. 32, no. 5, pp. 3875–3883, 2017, doi: 10.1109/TPWRS.2016.2632156.

[209] *IEEE standard for synchrophasor data transfer for power systems*, C37.118.2-2011.

[210] A. G. Phadke *et al.*, "Synchronized sampling and phasor measurements for relaying and control," *IEEE Trans. Power Delivery*, vol. 9, no. 1, pp. 442–452, 1994, doi: 10.1109/61.277716.

[211] *IEEE standard for synchrophasor measurements for power systems*, C37.118.1-2011.

[212] *Measuring relays and protection equipment – Part 118-1: Synchrophasor for power systems – Measurements*, IEC/IEEE 60255-118-1, 2018.

[213] *IEEE guide for phasor data concentrator requirements for power system protection, control, and monitoring*, IEEE Std C37.244-2013.

[214] A. Monti, C. Muscas, and F. Ponci, Eds., *Phasor measurement units and wide area monitoring systems: From the sensors to the system*, 2016. [Online]. Available: http://www.sciencedirect.com/science/book/9780128045695

[215] *IEEE Guide for Synchronization, Calibration, Testing, and Installation of Phasor Measurement Units (PMUs) for Power System Protection and Control*, IEEE C37.242-2013, 2013.

[216] Cigré WG C2.17, "Wide area monitoring systems – Support for control room applications," Technical Brochure 750, 2018.

[217] A. G. Phadke and J. S. Thorp, *Synchronized Phasor Measurements and Their Applications,* 2nd ed. Cham, s.l.: Springer International Publishing, 2017. [Online]. Available: http://dx.doi.org/10.1007/978-3-319-50584-8

[218] Y. Liu, L. Wu, and J. Li, "D-PMU based applications for emerging active distribution systems: A review," *Electric Power Systems Research*, vol. 179, p. 106063, 2020, doi: 10.1016/j.epsr.2019.106063.

[219] J. de La Ree, V. Centeno, J. S. Thorp, and A. G. Phadke, "Synchronized Phasor Measurement Applications in Power Systems," *IEEE Trans. Smart Grid*, vol. 1, no. 1, pp. 20–27, 2010, doi: 10.1109/TSG.2010.2044815.

[220] Seyed Younes Seyedi, "Synchrophasor data analytics for control and protection applications in smart grids," PhD Dissertation, École polytechnique de Montréal, Université de Montreál, Montreal, 2017.

[221] NASPI, "Synchrophasor starter kit: Draft October 9, 2015," North American Synchrophasor Initiative (NASPI). Accessed: Mar. 21 2020. [Online]. Available: https://www.naspi.org/sites/default/files/reference_documents/4.pdf

[222] A. R. Messina, *Wide-area monitoring of interconnected power systems*. London: The Institution of Engineering and Technology, 2015.

[223] *Wide area monitoring and control for transmission capability enhancement*. Paris: Cigré, 2007.

[224] M. Parashar, A. Jampala, and J. Giri, "Practical Experiences in Operationalizing WAMS in the Control Room," *IFAC-PapersOnLine*, vol. 49, no. 27, pp. 91–95, 2016, doi: 10.1016/j.ifacol.2016.10.725.

[225] A. G. Phadke *et al.*, "The Wide World of Wide-area Measurement," *IEEE Power and Energy Mag.*, vol. 6, no. 5, pp. 52–65, 2008, doi: 10.1109/MPE.2008.927476.

[226] W. Sattinger and G. Giannuzzi, "Monitoring Continental Europe: An Overview of WAM Systems Used in Italy and Switzerland," *IEEE Power and Energy Mag.*, vol. 13, no. 5, pp. 41–48, 2015, doi: 10.1109/MPE.2015.2431215.

[227] Y. Zhang and A. Bose, "Design of Wide-Area Damping Controllers for Interarea Oscillations," *IEEE Trans. Power Syst.*, vol. 23, no. 3, pp. 1136–1143, 2008, doi: 10.1109/TPWRS.2008.926718.

[228] R. Preece, J. V. Milanovic, A. M. Almutairi, and O. Marjanovic, "Damping of inter-area oscillations in mixed AC/DC networks using WAMS based supplementary controller," *IEEE Trans. Power Syst.*, vol. 28, no. 2, pp. 1160–1169, 2013, doi: 10.1109/TPWRS.2012.2207745.

[229] C. W. Taylor, D. C. Erickson, K. E. Martin, R. E. Wilson, and V. Venkatasubramanian, "WACS-Wide-Area Stability and Voltage Control System: R&D and Online Demonstration," *Proc. IEEE*, vol. 93, no. 5, pp. 892–906, 2005, doi: 10.1109/JPROC.2005.846338.

[230] Cigré WG C2.17, "Wide Area protection and Control Technologies," Technical Brochure 664, 2016.

[231] V. Terzija *et al.*, "Wide-Area Monitoring, Protection, and Control of Future Electric Power Networks," *Proc. IEEE*, vol. 99, no. 1, pp. 80–93, 2011, doi: 10.1109/JPROC.2010.2060450.

[232] X. Xie, Y. Xin, J. Xiao, J. Wu, and Y. Han, "WAMS applications in Chinese power systems," *IEEE Power and Energy Mag.*, vol. 4, no. 1, pp. 54–63, 2006, doi: 10.1109/MPAE.2006.1578532.

[233] P. Wall *et al.*, "Deployment and demonstration of wide area monitoring system in power system of Great Britain," *J. Mod. Power Syst. Clean Energy*, vol. 4, no. 3, pp. 506–518, 2016, doi: 10.1007/s40565-016-0218-3.

[234] M. Zima, M. Larsson, P. Korba, C. Rehtanz, and G. Andersson, "Design Aspects for Wide-Area Monitoring and Control Systems," *Proc. IEEE*, vol. 93, no. 5, pp. 980–996, 2005, doi: 10.1109/JPROC.2005.846336.

[235] A. G. Phadke and T. Bi, "Phasor measurement units, WAMS, and their applications in protection and control of power systems," *J. Mod. Power Syst. Clean Energy*, vol. 6, no. 4, pp. 619–629, 2018, doi: 10.1007/s40565-018-0423-3.

[236] K. Tuttelberg, J. Kilter, D. Wilson, and K. Uhlen, "Estimation of Power System Inertia From Ambient Wide Area Measurements," *IEEE Trans. Power Syst.*, vol. 33, no. 6, pp. 7249–7257, 2018, doi: 10.1109/TPWRS.2018.2843381.

[237] Cigré WG C4.34, "Application of phasor measurement units for monitoring powersystem dynamic performance," Technical Brochure 702, 2017.

[238] J. Tang, J. Liu, F. Ponci, and A. Monti, "Adaptive load shedding based on combined frequency and voltage stability assessment using synchrophasor measurements," *IEEE Trans. Power Syst.*, vol. 28, no. 2, pp. 2035–2047, 2013, doi: 10.1109/TPWRS.2013.2241794.

[239] Adamiak, Novosel, Kasztenny, Madani, Sykes, and Phadke, "Wide Area Protection and Control - Today and Tomorrow," in *2005/2006 IEEE/PES Transmission and Distribution Conference and Exhibition*, Dallas, TX, 2006.

[240] B. P. F. Wu, "The application of expert systems to process control," *JOM*, vol. 43, no. 1, pp. 13–16, 1991, doi: 10.1007/BF03220111.

[241] P. N. Finlay, "Decision support systems and expert systems: A comparison of their components and design methodologies," *Computers & Operations Research*, vol. 17, no. 6, pp. 535–543, 1990, doi: 10.1016/0305-0548(90)90059-G.

[242] D. Niebur, "Expert systems for power system control in Western Europe," *IEEE Control Syst.*, vol. 11, no. 4, pp. 34–40, 1991, doi: 10.1109/37.88588.

[243] Tharam S. Dillon and Michael Laughton, Eds., *Expert System Applications in Power Systems*: Prentice Hall International (UK) Ltd., 1990.

[244] R. Parasuraman, T. B. Sheridan, and C. D. Wickens, "A model for types and levels of human interaction with automation," *IEEE transactions on systems, man, and cybernetics. Part A, Systems and*

humans : a publication of the IEEE Systems, Man, and Cybernetics Society, vol. 30, no. 3, pp. 286–297, 2000, doi: 10.1109/3468.844354.

[245] K. P. Wong, "Applications of Artificial Intelligence and Expert Systems in Power Engineering," in *Expert Systems in Engineering Applications*, S. Tzafestas, Ed., Berlin, Heidelberg: Springer Berlin Heidelberg, 1993, pp. 160–178.

[246] M. J. E. Powell, "Data acquisition and requirements for expert systems in a control room context (power systems)," in *IEE Colloquium on Expert Systems in the Field of Protection and Control*, 1993, 3/1-3/2.

[247] G. Kreuzberger, A. Lunzer, and R. Kaschek, *Interdisciplinary Advances in Adaptive and Intelligent Assistant Systems*: IGI Global, 2011.

[248] A. Marot, A. Rozier, M. Dussartre, L. Crochepierre, and B. Donnot, "Towards an AI Assistant for Power Grid Operators," in *Frontiers in Artificial Intelligence and Applications, HHAI2022: Augmenting Human Intellect*, S. Schlobach, M. Pérez-Ortiz, and M. Tielman, Eds.: IOS Press, 2022.

[249] A. Kubis, J. Hachenberger, M. Heine, J. Eisele, T. Dong, T. Rzymek, M. Abel, M. Thiele, H. Woiton, R. Fuchs, A. Gumbel, "Online Security Assessment and System Optimization for Close to Real-Time Decision Support: Recent Advances and Lessons Learned from a Joint Development Project," in *Cigré eSession 2020*.

[250] M. Heine and B. Jockschat, "SASO – On the way to auto pilot in network control systems," in *VDE-Kongress 2016: Internet der Dinge : Technologien, Anwendungen, Perspektiven : 07./08.11.2016, Center Mannheim : Kongressbeiträge*, Berlin: VDE Verlag, 2016.

[251] F. Sass, T. Sennewald, C. Brosinsky, M. Mangold, C. Heyde, S. Becher, R. Krebs and D. Westermann, "Control Center Implementation of Advanced Optimization and Decision Support Applications," in *International Conference on Smart Energy Systems and Technologies (SEST), Sevilla*, 2018.

[252] Laure Crochepierre, "Apprentissage automatique interactif pour les opérateurs du réseau électrique," Thesé pour l'obtention du Doctorat, Université de Lorraine, Lorraine, France, 2022. Accessed: Nov. 12 2022. [Online]. Available: http://docnum.univ-lorraine.fr/public/DDOC_T_2022_0112_CROCHEPIERRE.pdf

[253] J. Yang, W. Li, T. Chen, W. Xu, and M. Wu, "Online estimation and application of power grid impedance matrices based on synchronised phasor measurements," *IET Generation, Transmission & Distribution*, vol. 4, no. 9, p. 1052, 2010, doi: 10.1049/iet-gtd.2010.0021.

[254] D. Ritzmann, P. S. Wright, W. Holderbaum, and B. Potter, "A Method for Accurate Transmission Line Impedance Parameter Estimation," *IEEE Trans. Instrum. Meas.*, vol. 65, no. 10, pp. 2204–2213, 2016, doi: 10.1109/TIM.2016.2556920.

[255] Y. Du and Y. Liao, "On-line estimation of transmission line parameters, temperature and sag using PMU measurements," *Electric Power Systems Research*, vol. 93, pp. 39–45, 2012, doi: 10.1016/j.epsr.2012.07.007.

[256] S. S. Mousavi-Seyedi, F. Aminifar, and S. Afsharnia, "Parameter Estimation of Multiterminal Transmission Lines Using Joint PMU and SCADA Data," *IEEE Trans. Power Delivery*, vol. 30, no. 3, pp. 1077–1085, 2015, doi: 10.1109/TPWRD.2014.2369500.

[257] M. Asprou and E. Kyriakides, "Identification and Estimation of Erroneous Transmission Line Parameters Using PMU Measurements," *IEEE Trans. Power Delivery*, p. 1, 2017, doi: 10.1109/TPWRD.2017.2648881.

[258] J. Wibbeke, M. Aldebs, D. Babazadeh, P. Teimourzadeh Baboli, and S. Lehnhoff, "Power Systems Digital Twin under Measurement and Model Uncertainties: Network Parameter Tuning Approach," in *2021 IEEE Madrid PowerTech*, Madrid, Spain, 2021.

[259] I. D. Landau and A. Karimi, "A recursive algorithm for ARMAX model identification in closed loop," *IEEE Trans. Automat. Contr.*, vol. 44, no. 4, pp. 840–843, 1999, doi: 10.1109/9.754830.

[260] L. Fan, Z. Miao, and Y. Wehbe, "Application of Dynamic State and Parameter Estimation Techniques on Real-World Data," *IEEE Trans. Smart Grid*, vol. 4, no. 2, pp. 1133–1141, 2013, doi: 10.1109/TSG.2012.2230031.

[261] A. Rouhani and A. Abur, "Constrained Iterated Unscented Kalman Filter for Dynamic State and Parameter Estimation," *IEEE Trans. Power Syst.*, vol. 33, no. 3, pp. 2404–2414, 2018, doi: 10.1109/TPWRS.2017.2764005.

[262] G. Wang, S.-J. Kim, and G. B. Giannakis, "Moving-horizon dynamic power system state estimation using semidefinite relaxation," in *2014 IEEE PES General Meeting | Conference & Exposition*, National Harbor, MD, USA, 2014.

[263] T. Chen, T. Tran, D. Zhou, W.-K. Ho, and K.-V. Ling, "Iterated moving horizon estimation for power systems," in *The 2014 International Conference on Control, Automation and Information Sciences (ICCAIS 2014)*, Gwangju, South Korea, 2014, pp. 81–86.

[264] T. Chen, "Robust state estimation for power systems via moving horizon strategy," *Sustainable Energy, Grids and Networks*, vol. 10, pp. 46–54, 2017, doi: 10.1016/j.segan.2017.02.005.

[265] A. Venzke, "Machine Learning and Convex Optimization for Secure Power System Operation," PhD Dissertation, 2020. Accessed: Sep. 21 2021. [Online]. Available: https://backend.orbit.dtu.dk/ws/portalfiles/portal/235697430/Thesis_Andreas_Horst_Venzke.pdf

[266] J. L. Cremer, "Probabilistic dynamic security assessment for power system control," PhD Dissertation, Department of Electrical and Electronic Engineering, Imperial College London, 2020.

[267] H. Cui, F. Li, and K. Tomsovic, "Cyber-physical system testbed for power system monitoring and wide-area control verification," *IET Energy Systems Integration*, vol. 93, no. 5, p. 965, 2019, doi: 10.1049/iet-esi.2019.0084.

[268] T. Becejac, C. Eppinger, A. Ashok, U. Agrawal, and J. OBrien, "PRIME: A real-time cyber-physical systems testbed: From Wide-Area Monitoring, Protection and Control (WAMPAC) prototyping to operator training and beyond," *IET Cyber-Physical Systems: Theory & Applications*, 2020, doi: 10.1049/iet-cps.2019.0049.

[269] L. Shi, Q. Dai, and Y. Ni, "Cyber–physical interactions in power systems: A review of models, methods, and applications," *Electric Power Systems Research*, vol. 163, pp. 396–412, 2018, doi: 10.1016/j.epsr.2018.07.015.

[270] S. Xin, Q. Guo, H. Sun, B. Zhang, J. Wang, and C. Chen, "Cyber-Physical Modeling and Cyber-Contingency Assessment of Hierarchical Control Systems," *IEEE Trans. Smart Grid*, vol. 6, no. 5, pp. 2375–2385, 2015, doi: 10.1109/TSG.2014.2387381.

[271] S. Zonouz, C. M. Davis, K. R. Davis, R. Berthier, R. B. Bobba, and W. H. Sanders, "SOCCA: A Security-Oriented Cyber-Physical Contingency Analysis in Power Infrastructures," *IEEE Trans. Smart Grid*, vol. 5, no. 1, pp. 3–13, 2014, doi: 10.1109/TSG.2013.2280399.

[272] Bob Piascik, John Vickers, and et al., *DRAFT Materials, Structures, Mechanical Systems, and Manufacturing Roadmap*. Technology Area 12. [Online]. Available: https://www.nasa.gov/pdf/501625main_TA12-MSMSM-DRAFT-Nov2010-A.pdf (accessed: Oct. 24 2022).

[273] Michael Grieves, "Origins of the Digital Twin Concept," Working Paper, Florida Institute of Technology, 2016. Accessed: Sep. 21 2021. [Online]. Available: https://www.researchgate.net/publication/307509727_Origins_of_the_Digital_Twin_Concept

[274] M. Grieves and J. Vickers, "Digital Twin: Mitigating Unpredictable, Undesirable Emergent Behavior in Complex Systems," in *Transdisciplinary Perspectives on Complex Systems: New Findings and Approaches*, F.-J. Kahlen, S. Flumerfelt, and A. Alves, Eds., Cham, s.l.: Springer International Publishing, 2017, pp. 85–113.

[275] J.-F. Uhlenkamp, K. Hribernik, S. Wellsandt, and K.-D. Thoben, "Digital Twin Applications: A first systemization of their dimensions," in *2019 IEEE International Conference on Engineering, Technology and Innovation (ICE/ITMC)*, 2019.

[276] K. Panetta, *Gartner Top 10 Strategic Technology Trends for 2018.* [Online]. Available: https://www.gartner.com/smarterwithgartner/gartner-top-10-strategic-technology-trends-for-2018/ (accessed: Apr. 5 2019).

[277] T. H.-J. Uhlemann, C. Lehmann, and R. Steinhilper, "The Digital Twin: Realizing the Cyber-Physical Production System for Industry 4.0," *Procedia CIRP*, vol. 61, pp. 335–340, 2017, doi: 10.1016/j.procir.2016.11.152.

[278] T. H.-J. Uhlemann, C. Schock, C. Lehmann, S. Freiberger, and R. Steinhilper, "The Digital Twin: Demonstrating the Potential of Real Time Data Acquisition in Production Systems," *Procedia Manufacturing*, vol. 9, pp. 113–120, 2017, doi: 10.1016/j.promfg.2017.04.043.

[279] B. Ashtari Talkhestani *et al.*, "An architecture of an Intelligent Digital Twin in a Cyber-Physical Production System," *at - Automatisierungstechnik*, vol. 67, no. 9, pp. 762–782, 2019, doi: 10.1515/auto-2019-0039.

[280] R. Rosen, G. von Wichert, G. Lo, and K. D. Bettenhausen, "About The Importance of Autonomy and Digital Twins for the Future of Manufacturing," *IFAC-PapersOnLine*, vol. 48, no. 3, pp. 567–572, 2015, doi: 10.1016/j.ifacol.2015.06.141.

[281] R. Stark, S. Kind, and S. Neumeyer, "Innovations in digital modelling for next generation manufacturing system design," *CIRP Annals*, vol. 66, no. 1, pp. 169–172, 2017, doi: 10.1016/j.cirp.2017.04.045.

[282] C. Brosinsky, M. Karaçelebi, and J. L. Cremer, "Machine Learning and Digital Twins: Monitoring and control for dynamic security," in *Monitoring and Control of Electrical Power Systems using Machine Learning Techniques*, Emilio Barocio Espejo, Felix Rafael Segundo Sevilla, Petr Korba, Ed.: Elsevier, 2023.

[283] B. Tekinerdogan and C. Verdouw, "Systems Architecture Design Pattern Catalogfor Developing Digital Twins," *Sensors (Basel, Switzerland)*, vol. 20, no. 18, 2020, doi: 10.3390/s20185103.

[284] A.-K. Marten, V. Akhmatov, T. Sørensen, R. Stornowski, D. Westermann, and C. Brosinsky, "Kriegers Flak - Combined Grid Solution: Coordinated Cross-Border Control of a Meshed HVAC/HVDC Offshore Wind Power Grid," *IET Renewable Power Generation*, vol. 12, no. 13, pp. 1493–1499, 2018, doi: 10.1049/iet-rpg.2017.0792.

[285] *Virtual commissioning - Model types and glossary*, VDI/VDE 3693 Blatt 1, 2016.

[286] E. VanDerHorn and S. Mahadevan, "Digital Twin: Generalization, characterization and implementation," *Decision Support Systems*, vol. 145, p. 113524, 2021, doi: 10.1016/j.dss.2021.113524.

[287] C. Brosinsky, A. Kummerow, M. Richter, A. Naumann, P. Wiest, S. Nicolai, and D. Westermann, "The Role of Digital Twins in Power System Automation and Control: Necessity, Requirements, and Benefits," in *ETG-Fachberichte, ETG-Kongress: Das Gesamtsystem im Fokus der Energiewende 18. - 19. Mai 2021, Online-Veranstaltung,* 1st ed., Berlin: VDE Verlag, 2021.

[288] *Assimilation for Atmospheric, Oceanic and Hydrologic Applications (Vol. II)*: Springer Berlin Heidelberg, 2013.

[289] K.-J. Wang, T.-L. Lee, and Y. Hsu, "Revolution on digital twin technology—a patent research approach," *Int J Adv Manuf Technol*, vol. 107, 11-12, pp. 4687–4704, 2020, doi: 10.1007/s00170-020-05314-w.

[290] Boschert, S., Rosen R., Heinrich, C., "Next generation digital Twin," in *Tools and methods of competitive engineering: Proceedings of the Twelfth International Symposium on Tools and Methods of Competitive Engineering - TMCE 2018*, I. Horváth, J. P. Suárez Rivero, and P. M. Hernández Castellano, Eds.: Delft University of Technology, 2018, pp. 209–218.

[291] C. Wagner *et al.,* "The role of the Industry 4.0 asset administration shell and the digital twin during the life cycle of a plant," in *2017 22nd IEEE International Conference on Emerging Technologies and Factory Automation (ETFA)*, Limassol, 2017.

[292] R. Heidel, M. Hoffmeister, and M. Hankel, *Industrie 4.0: The Reference Architecture Model RAMI 4.0 and the Industrie 4.0 component,* 1st ed., 2019.

[293] R. Saracco, "Digital Twins: Bridging Physical Space and Cyberspace," *Computer*, vol. 52, no. 12, pp. 58–64, 2019, doi: 10.1109/MC.2019.2942803.

[294] *Industrial-process measurement, control and automation - Life-cycle-management for systems and components*, IEC 62890:2020, 2020.

[295] C. Gehrmann and M. Gunnarsson, "A Digital Twin Based Industrial Automation and Control System Security Architecture," *IEEE Trans. Ind. Inf.*, vol. 16, no. 1, pp. 669–680, 2020, doi: 10.1109/TII.2019.2938885.

[296] G. Steindl, M. Stagl, L. Kasper, W. Kastner, and R. Hofmann, "Generic Digital Twin Architecture for Industrial Energy Systems," *Applied Sciences*, vol. 10, no. 24, p. 8903, 2020, doi: 10.3390/app10248903.

[297] B. Klenz, *How to Use Streaming Analytics to Create a Real-Time Digital Twin.* [Online]. Available: https://www.sas.com/content/dam/SAS/support/en/sas-global-forum-proceedings/2018/2004-2018.pdf (accessed: Apr. 5 2019).

[298] GE Power Digital Solutions, *GE Digital Twin: Analytic Engine for the Digital Power Plant.* [Online]. Available: https://www.ge.com/digital/sites/default/files/download_assets/Digital-Twin-for-the-digital-power-plant-.pdf (accessed: Aug. 20 2019).

[299] Aaron Parrot and Lane Warshaw, "Industry 4.0 and digital twin: Manufacturing meets its match," *Deloitte University Press*, 2017.

[300] MarketsandMarkets, *Digital Twin Market worth $48.2 billion by 2026.* [Online]. Available: https://www.marketsandmarkets.com/Market-Reports/digital-twin-market-225269522.html (accessed: Aug. 27 2021).

[301] S. Boschert and R. Rosen, "Digital Twin—The Simulation Aspect," in *Mechatronic Futures*, P. Hehenberger and D. Bradley, Eds., Cham: Springer International Publishing, 2016, pp. 59–74.

[302] A. El Saddik, "Digital Twins: The Convergence of Multimedia Technologies," *IEEE MultiMedia*, vol. 25, no. 2, pp. 87–92, 2018, doi: 10.1109/MMUL.2018.023121167.

[303] Cigré Task Force 38.02.19, "System protection schemes in power networks," Cigré Technical Brochure 187, Jun. 2001.

[304] Cigré Working Group C2.02.24, "Defense Plan against extreme contingencies," ELECTRA 231, Apr. 2007.

[305] V. Madani, M. Parashar, A. Jampala, and J. Giri, "Frontiers of synchrophasor solutions deployment," in *2017 7th International Conference on Power Systems (ICPS): College of Engineering Pune, India, Dec 21-23, 2017*, Pune, 2017, pp. 514–520.

[306] P. Palensky, M. Cvetkovic, D. Gusain, and A. Joseph, "Digital twins and their use in future power systems," *digitaltwin*, vol. 1, p. 4, 2021, doi: 10.12688/digitaltwin.17435.1.

[307] S. P. Kovalyov, "Design and Development of a Power System Digital Twin: A Model-based Approach," in *2021 3rd International Conference on Control Systems, Mathematical Modeling, Automation and Energy Efficiency (SUMMA)*, Lipetsk, Russian Federation, 2021, pp. 843–848.

[308] Siemens AG, *Siemens Electrical Digital Twin: A single source of truth to unlock the potential within a modern utility's data landscape.* [Online]. Available: siemens.com/electrical-digital-twin (accessed: Jan. 22 2021).

[309] *Enterprise-control system integration - Part 1 to 6*, IEC 62264, 2007.

[310] A. Roth, *Einführung und Umsetzung von Industrie 4.0.* Berlin, Heidelberg: Springer Berlin Heidelberg, 2016.

[311] John Erik Hershey, Frederick Wilson Wheeler, Matthew Christian Nielsen, Christopher Donald Johnson, Michael Joseph Dell'Anno, and Joij Joykutti, "Digital twin of twinned physical system," US 2017/0286572 A1.

[312] ABB, "Digital twins and simulations," Feb. 2019. Accessed: 01.204.2020. [Online]. Available: https://search.abb.com/library/Download.aspx?DocumentID=9AKK107492A3437&LanguageCode=en&DocumentPartId=&Action=Launch

[313] M. Mirz, "A dynamic phasor real-time simulation based digital twin for power systems," Dissertation, RWTH Aachen University, Aachen, 2020. [Online]. Available: https://publications.rwth-aachen.de/record/804608

[314] Q. D. Vu, "Parameter estimation in complex nonlinear dynamical systems," Dissertation, Technische Universität Ilmenau, Ilmenau, 2015. [Online]. Available: http://www.db-thueringen.de/servlets/DocumentServlet?id=26727

[315] M. Rau, "Nichtlineare modellbasierte prädiktive Regelung auf Basis lernfähiger Zustandsraummodelle," Dissertation, Fakultät für Elektrotechnik und Informationstechnik, Technische Universität München, München, 2003.

[316] L. Ljung, "Perspectives on system identification," *Annual Reviews in Control*, vol. 34, no. 1, pp. 1–12, 2010, doi: 10.1016/j.arcontrol.2009.12.001.

[317] K. E. Bollinger, H. S. Khalil, L. Li, and W. Norum, "A Method for on-Line Identification of Power System Model Parameters in the Presence of Noise," *IEEE Trans. on Power Apparatus and Syst.*, PAS-101, no. 9, pp. 3105–3111, 1982, doi: 10.1109/TPAS.1982.317522.

[318] J. Hachenberger *et al.*, "Development of a Preventive and Curative Congestion Management Module for Close to Real-Time Transmission System Operations," in *ETG Congress 2021*, 2021, pp. 1–6.

[319] O. Nelles, *Nonlinear System Identification: From Classical Approaches to Neural Networks and Fuzzy Models*. Berlin, Heidelberg: Springer, 2001. [Online]. Available: http://dx.doi.org/10.1007/978-3-662-04323-3

[320] J. Schoukens and L. Ljung, "Nonlinear System Identification: A User-Oriented Road Map," *IEEE Control Systems Magazine*, vol. 39, no. 6, pp. 28–99, 2019.

[321] P. M. Pardalos, D. Hearn, and K. Schittkowski, *Numerical Data Fitting in Dynamical Systems*. Boston, MA: Springer US, 2002.

[322] M. Diehl, H. J. Ferreau, and N. Haverbeke, "Efficient Numerical Methods for Nonlinear MPC and Moving Horizon Estimation," *Lecture Notes in Control and Information Sciences*, pp. 391–417, 2009.

[323] V. Kleyman *et al., State and parameter estimation for model-based retinal laser treatment.* [Online]. Available: arxiv://2103.03189

[324] *Communication networks and systems for power utility automation - Part 90-2: Using IEC 61850 for communication between substations and control centres*, IEC TR 61850-90-2:2016, 2016.

[325] K. Zhu, *Data quality in wide-area monitoring and control systems: PMU data latency, completness and design of wide-area damping systems*. Stockholm: KTH Royal Institute of Technology, 2013.

[326] A. E. Saldaña-González, A. Sumper, M. Aragüés-Peñalba, and M. Smolnikar, "Advanced Distribution Measurement Technologies and Data Applications for Smart Grids: A Review," *Energies*, vol. 13, no. 14, p. 3730, 2020, doi: 10.3390/en13143730.

[327] M. Brown, M. Biswal, S. Brahma, S. J. Ranade, and H. Cao, Eds., *Characterizing and quantifying noise in PMU data*. 2016 IEEE Power and Energy Society General Meeting (PESGM), 2016.

[328] B. Chaudhuri, R. Majumder, and B. C. Pal, "Wide-Area Measurement-Based Stabilizing Control of Power System Considering Signal Transmission Delay," *IEEE Trans. Power Syst.*, vol. 19, no. 4, pp. 1971–1979, 2004, doi: 10.1109/TPWRS.2004.835669.

[329] K. Zhu, L. Nordstrom, and A. T. Al-Hammouri, "Examination of data delay and packet loss for wide-area monitoring and control systems," in *2012 IEEE International Energy Conference and Exhibition (ENERGYCON 2012): Florence, Italy, 9 - 12 September 2012*, Florence, Italy, 2012, pp. 927–934.

[330] P. Kansal and A. Bose, "Bandwidth and Latency Requirements for Smart Transmission Grid Applications," *IEEE Trans. Smart Grid*, vol. 3, no. 3, pp. 1344–1352, 2012, doi: 10.1109/TSG.2012.2197229.

[331] Substations Committee of the IEEE Power Engineering Society, *IEEE recommended practice for data communications between remote terminal units and intelligent electronic devices in a substation.* New York: Institute of Electrical and Electronics Engineers, 2001. [Online]. Available: http://ieeexplore.ieee.org/servlet/opac?punumber=7287

[332] *IEEE Standard Communication Delivery Time Performance Requirements for Electric Power Substation Automation*, IEEE 1646, 2004.

[333] J. Wu and N. N. Schulz, "Overview of CIM-oriented database design and data exchanging in power system applications," in *Proceedings of the 37th Annual North American Power Symposium, 2005*, Ames, IA, USA, 2005, pp. 16–20.

[334] A. Sharma, S. C. Srivastava, and S. Chakrabarti, "An Extension of Common Information Model for Power System Multiarea State Estimation," *IEEE Systems Journal*, vol. 11, no. 3, pp. 1692–1701, 2017, doi: 10.1109/JSYST.2014.2372813.

[335] G. Ravikumar and S. A. Khaparde, "A Common Information Model Oriented Graph Database Framework for Power Systems," *IEEE Trans. Power Syst.*, vol. 32, no. 4, pp. 2560–2569, 2017, doi: 10.1109/TPWRS.2016.2631242.

[336] ISO/IEC 2382:2015, *Information technology — Vocabulary.* [Online]. Available: https://www.iso.org/obp/ui/ (accessed: Nov. 26 2018).

[337] A. G. Phadke and B. Kasztenny, "Synchronized Phasor and Frequency Measurement Under Transient Conditions," *IEEE Trans. Power Delivery*, vol. 24, no. 1, pp. 89–95, 2009, doi: 10.1109/TPWRD.2008.2002665.

[338] J. Machowski, J. W. Bialek, and J. R. Bumby, *Power system dynamics: Stability and control,* 2nd ed. Chichester: Wiley, 2012.

[339] S. Henschel, "Analysis of electromagnetic and electromechanical power system transients with dynamic phasors," PhD Dissertation, The University of British Columbia, 1999.

[340] M. Amin, "National Infrastructures as Complex Interactive Networks," in *Automation, Control, and Complexity: An Integrated Approach*, Samad & Weyrauch, Ed.: John Wiley and Sons, 2000, pp. 263–286.

[341] S. Cole, "Steady-state and dynamic modelling of VSC HVDC systems for power system simulation," PhD Dissertation, Faculteit Ingenieurswetenschappen, Katholieke Universiteit Leuven, Leuven, 2010.

[342] L. O. Chua and P.-M. Lin, *Computer-aided analysis of electronic circuits: Algorithms and computational techniques*. Englewood Cliffs, NJ: Prentice-Hall, 1975.

[343] W. H. Press, *Numerical recipes: The art of scientific computing,* 3rd ed. Cambridge: Cambridge Univ. Press, 2007.

[344] J. C. Butcher, *Numerical methods for ordinary differential equations,* 3rd ed. Chichester, UK: Wiley, 2016.

[345] S. Jin, Z. Huang, R. Diao, Di Wu, and Y. Chen, "Comparative Implementation of High Performance Computing for Power System Dynamic Simulations," *IEEE Trans. Smart Grid*, vol. 8, no. 3, pp. 1387–1395, 2017, doi: 10.1109/TSG.2016.2647220.

[346] S. Cole and R. Belmans, "MatDyn, A New Matlab-Based Toolbox for Power System Dynamic Simulation," *IEEE Trans. Power Syst.*, vol. 26, no. 3, pp. 1129–1136, 2011, doi: 10.1109/TPWRS.2010.2071888.

[347] F. Milano, *Power System Modelling and Scripting*. Berlin, Heidelberg: Springer Berlin Heidelberg, 2010.

[348] V. Crastan and D. Westermann, Eds., *Elektrische Energieversorgung 3: Dynamik, Regelung und Stabilität, Versorgungsqualität, Netzplanung, Betriebsplanung und -führung, Leit- und Informationstechnik, FACTS, HGÜ,* 2nd ed. Berlin: Springer Vieweg, 2018. [Online]. Available: http://dx.doi.org/10.1007/978-3-662-49021-1

[349] Z. Huang *et al.*, "Capturing real-time power system dynamics: Opportunities and challenges," in *2015 IEEE Power & Energy Society General Meeting*, Denver, CO, USA, 2015, pp. 1–5.

[350] T. Park and P. I. Barton, "State event location in differential-algebraic models," *ACM Trans. Model. Comput. Simul.*, vol. 6, no. 2, pp. 137–165, 1996, doi: 10.1145/232807.232809.

[351] DIgSILENT GmbH, *DIgSILENT PowerFactory User Manual: Version 2019*. Online Edition, 2018.

[352] *Middleware in industrial automation - Fundamentals*, VDI/VDE 2657 Blatt 1, 2013.

[353] Apache Kafka, *A Distributed Streaming Platform*. [Online]. Available: https://kafka.apache.org/ (accessed: Nov. 2 2020).

[354] B. Stopford and S. Newman, *Designing Event-Driven Systems: Concepts and Patterns for Streaming Services with Apache Kafka*. Beijing, Boston, Farnham, Sebastopol, Tokyo: O'Really Media, 2018.

[355] C. Scheibe *et al.*, "A Novel Co-Simulation Concept using Interprocess Communication in Shared Memory," in *2019 IEEE Power & Energy Society General Meeting (PESGM)*, Atlanta, GA, USA, 2019.

[356] M. Oettmeier, "Weiterentwicklung einer dynamischen Netzberechnungssoftware zur simultanen Co-Simulation in mehreren Instanzen unter Verwendung eines standardisierten Datenaustauschverfahrens," Master Thesis, Fachgebiet Elektrische Energieversorgung, Technische Universität Ilmenau, Ilmenau, 2019.

[357] P. L. Dandeno, "Current Usage & Suggested Practices in Power System Stability Simulations for Synchronous Machines," *IEEE Trans. Energy Convers.*, EC-1, no. 1, pp. 77–93, 1986, doi: 10.1109/TEC.1986.4765673.

[358] K. R. Padiyar, *Power system dynamics: Stability and control,* 2nd ed. Hyderabad [India]: BS Publications, 2008.

[359] IEEE Standards Association, *IEEE Guide for Synchronous Generator Modeling Practices and Parameter Verification with Applications in Power System Stability Analyses.*

[360] T. Weckesser, H. Johannsson, and J. Ostergaard, "Impact of model detail of synchronous machines on real-time transient stability assessment," in *2013 IREP Symposium Bulk Power System Dynamics and Control - IX Optimization, Security and Control of the Emerging Power Grid*, Rethymno, Greece, 2013, pp. 1–9.

[361] I. M. Canay, "Determination of model parameters of synchronous machines," *IEE Proc. B Electr. Power Appl. UK*, vol. 130, no. 2, p. 86, 1983, doi: 10.1049/ip-b.1983.0014.

[362] Cigré Working Group B4.57, "Guide for the Development of Models for HVDC Converters in a HVDC Grid," Technical Brochure 604, 2014.

[363] S. Cole and R. Belmans, "A proposal for standard VSC HVDC dynamic models in power system stability studies," *Electric Power Systems Research*, vol. 81, no. 4, pp. 967–973, 2011, doi: 10.1016/j.epsr.2010.11.032.

[364] S. Wenig, "Potential of Bipolar Full-Bridge MMC-HVdc Transmission for Link and Overlay Grid Applications," Dissertation, Institut für Elektroenergiesysteme und Hochspannungstechnik (IEH), KIT, Karlsruhe, 2019. [Online]. Available: https://publikationen.bibliothek.kit.edu/1000096294

[365] N. R. Chaudhuri, A. Yazdani, B. Chaudhuri, and R. Majumder, *Multi-terminal direct-current grids: Modeling, analysis, and control.* Chichester: John Wiley & Sons, 2014. [Online]. Available: http://gbv.eblib.com/patron/FullRecord.aspx?p=4040187

[366] H. Latorre, *Modeling and control of VSC-HVDC transmissions*. Stockholm: Electrical Engineering, KTH Royal Institute of Technology, 2011.

[367] D. van Hertem, J. Liang, and O. Gomis-Bellmunt, Eds., *HVDC grids: For offshore and supergrid future*. Piscataway: IEEE Press Wiley, 2016.

[368] I. Jahn, *Protection for Multiterminal HVDC Grids - A Digital Contribution*. Stockholm: KTH Royal Institute of Technology, 2021.

[369] R. D. Zimmerman, C. E. Murillo-Sanchez, and R. J. Thomas, "MATPOWER: Steady-State Operations, Planning, and Analysis Tools for Power Systems Research and Education," *IEEE Trans. Power Syst.*, vol. 26, no. 1, pp. 12–19, 2011, doi: 10.1109/TPWRS.2010.2051168.

[370] D. Shi, D. J. Tylavsky, N. Logic, and K. M. Koellner, "Identification of short transmission-line parameters from synchrophasor measurements," in *40th North American Power Symposium, 2008: NAPS '08 ; 28 - 30 Sept. 2008, University of Calgary, Calgary, Alberta, Canada*, Calgary, AB, Canada, 2008, pp. 1–8.

[371] F. P. Albuquerque, E. C. M. Costa, L. H. B. Liboni, R. F. R. Pereira, and M. C. de Oliveira, "Estimation of transmission line parameters by using two least-squares methods," *IET Generation, Transmission & Distribution*, vol. 15, no. 3, pp. 568–575, 2021, doi: 10.1049/gtd2.12044.

[372] Y. Zhang and Y. Liao, "Kalman filter based method for tracking dynamic transmission line parameters," *SN Appl. Sci.*, vol. 3, no. 1, p. 43, 2021, doi: 10.1007/s42452-020-04036-8.

[373] W. W. Price, K. A. Wirgau, A. Murdoch, J. V. Mitsche, E. Vaahedi, and M. El-Kady, "Load modeling for power flow and transient stability computer studies," *IEEE Trans. Power Syst.*, vol. 3, no. 1, pp. 180–187, 1988, doi: 10.1109/59.43196.

[374] D. Karlsson and D. J. Hill, "Modelling and identification of nonlinear dynamic loads in power systems," *IEEE Trans. Power Syst.*, vol. 9, no. 1, pp. 157–166, 1994, doi: 10.1109/59.317546.

[375] J.-C. Wang, H.-D. Chiang, C.-L. Chang, A.-H. Liu, C.-H. Huang, and C.-Y. Huang, "Development of a frequency-dependent composite load model using the measurement approach," *IEEE Trans. Power Syst.*, vol. 9, no. 3, pp. 1546–1556, 1994, doi: 10.1109/59.336105.

[376] A. Rouhani, "Robust Dynamic State Estimation In Power Systems," PhD Dissertation, Department of Electrical and Computer Engineering, Northeastern University, Boston, USA, 2017.

[377] I. R. Navarro, "Dynamic Load Models for Power Systems: Estimation of Time-Varying Parameters During Normal Operation," Licentiate Thesis, Department of Industrial Electrical Engineering, Lund University, Lund, Sweden, 2002. Accessed: Nov. 8 2018. [Online]. Available: https://www.iea.lth.se /publications/Theses/LTH-IEA-1034.pdf

[378] J. V. Milanovic, K. Yamashita, S. Martinez Villanueva, S. Z. Djokic, and L. M. Korunovic, "International Industry Practice on Power System Load Modeling," *IEEE Trans. Power Syst.*, vol. 28, no. 3, pp. 3038–3046, 2013, doi: 10.1109/TPWRS.2012.2231969.

[379] "Load representation for dynamic performance analysis (of power systems)," *IEEE Trans. Power Syst.*, vol. 8, no. 2, pp. 472–482, 1993, doi: 10.1109/59.260837.

[380] B. R. Oswald, *Berechnung von Drehstromnetzen: Berechnung stationärer und nichtstationärer Vorgänge mit symmetrischen Komponenten und Raumzeigern,* 2nd ed. Wiesbaden: Springer Vieweg, 2013. [Online]. Available: http://dx.doi.org/10.1007/978-3-8348-2621-3

[381] M. Paolone *et al.,* "Fundamentals of power systems modelling in the presence of converter-interfaced generation," *Electric Power Systems Research,* vol. 189, p. 106811, 2020, doi: 10.1016/j.epsr.2020.106811.

[382] Cigré Joint Working Group C4/B4/C1.604, "Influence of Embedded HVDC Transmission on System Security and AC Network Performance," Technical Brochure 536, 2013.

[383] H. Ding, G. Trajcevski, P. Scheuermann, X. Wang, and E. Keogh, "Querying and mining of time series data," *Proc. VLDB Endow.*, vol. 1, no. 2, pp. 1542–1552, 2008, doi: 10.14778/1454159.1454226.

[384] M. Winkler and B. Oswald, "Ermittlung und Dokumentation der Modellparameter für die Simulation des nichtstationären Betriebsverhaltens von Elektroenergiesystemen," in *Wissenschaftliche Zeitschrift der Technischen*, Dresden, 1988, pp. 175–181.

[385] DIgSILENT GmbH, "Technical Reference Documentation PowerFactory 2019: Synchronous Machine," Jul. 2019.

[386] B. Tan, J. Zhao, M. Netto, V. Krishnan, V. Terzija, and Y. Zhang, "Power system inertia estimation: Review of methods and the impacts of converter-interfaced generations," *International Journal of Electrical Power & Energy Systems*, vol. 134, no. 4, p. 107362, 2022, doi: 10.1016/j.ijepes.2021.107362.

[387] F. Milano and A. Ortega, "Frequency Divider," *IEEE Trans. Power Syst.*, p. 1, 2016, doi: 10.1109/TPWRS.2016.2569563.

[388] G. Misyris, "Towards Zero-Inertia Power Systems: Stability Analysis, Control & Physics-Informed Neural Networks," 2021.

[389] L. F. Rivera, M. Jimenez, N. Villegas, G. Tamura, and H. A. Muller, "The Forging of Autonomic and Cooperating Digital Twins," *IEEE Internet Computing*, p. 1, 2021, doi: 10.1109/MIC.2021.3051902.

[390] D. W. Gao *et al.,* "Application of AI techniques in monitoring and operation of power systems," *Front. Energy*, vol. 13, no. 1, pp. 71–85, 2019, doi: 10.1007/s11708-018-0589-4.

[391] A. Marot *et al.,* "Perspectives on Future Power System Control Centers for Energy Transition," *J. Mod. Power Syst. Clean Energy*, vol. 10, no. 2, pp. 328–344, 2022, doi: 10.35833/MPCE.2021.000673.

[392] B. Huang and J. Wang, "Applications of Physics-Informed Neural Networks in Power Systems - A Review," *IEEE Trans. Power Syst.*, p. 1, 2022, doi: 10.1109/TPWRS.2022.3162473.

[393] S. Kovalyov, "An Approach to Develop a Generative Design Technology for Power Systems," in *Proceedings of the VIth International Workshop 'Critical Infrastructures: Contingency Management, Intelligent, Agent-Based, Cloud Computing and Cyber Security' (IWCI 2019)*, Baikalsk, Russia, 2019.

[394] A. Marot *et al., Learning to run a power network with trust.* [Online]. Available: https://arxiv.org/abs/2110.12908

[395] A. Marot *et al.,* "Learning to run a Power Network Challenge: A Retrospective Analysis," 2021. Accessed: Jan. 1 2022. [Online]. Available: https://arxiv.org/abs/2103.03104

[396] *Communication networks and systems for power utility automation - Part 7-2: Basic information and communication structure - Abstract communication service interface (ACSI)*, IEC 61850-7-2:2010, 2010.

[397] International Electrotechnical Commission (IEC), *Electropedia: The World's Online Electrotechnical Vocabulary*. IEC Glossary. [Online]. Available: https://www.electropedia.org/ (accessed: Apr. 15 2020).

[398] *Energy management system application program interface (EMS-API) - Part 301: Common information model (CIM) base*, IEC 61970-301:2016, 2016.

[399] A. Monticelli and F. Wu, "Network Observability: Identification of Observable Islands and Measurement Placement," *IEEE Trans. on Power Apparatus and Syst.*, PAS-104, no. 5, pp. 1035–1041, 1985, doi: 10.1109/TPAS.1985.323453.

[400] IEEE PES Industry Technical Support Task Force, "The Definition and Quantification of Resilience," PES Techn. Rep. PES-TR 65, Apr. 2018.

[401] M. R. Endsley, "Toward a Theory of Situation Awareness in Dynamic Systems," *Human Factors: The Journal of the Human Factors and Ergonomics Society*, vol. 37, no. 1, pp. 32–64, 1995, doi: 10.1518/001872095779049543.

[402] N. Burgess and J. A. King, "Navigation in Virtual Space: Psychological and Neural Aspects," in *International Encyclopedia of the Social & Behavioral Sciences*, N. J. Smelser and P. B. Baltes, Eds., Oxford: Pergamon, 2001, pp. 10417–10422. [Online]. Available: http://www.sciencedirect.com/science/article/pii/B0080430767035646

[403] V. Crastan, *Elektrische Energieversorgung,* 4th ed. Heidelberg: Springer Vieweg, 2015.

[404] R. H. Park, "Two-reaction theory of synchronous machines generalized method of analysis-part I," *Trans. Am. Inst. Electr. Eng.*, vol. 48, no. 3, pp. 716–727, 1929, doi: 10.1109/T-AIEE.1929.5055275.

[405] P. Kundur and P. Dandeno, "Implementation of Advanced Generator Models into Power System Stability Programs," *IEEE Trans. on Power Apparatus and Syst.*, PAS-102, no. 7, pp. 2047–2054, 1983, doi: 10.1109/TPAS.1983.318173.

[406] C. B. Moler, *Numerical computing with MATLAB*. Philadelphia, Pa.: Society for Industrial and Applied Mathematics (SIAM 3600 Market Street Floor 6 Philadelphia PA 19104), 2008. [Online]. Available: http://epubs.siam.org/ebooks/siam/other_titles_in_applied_mathematics/ot87

[407] W. D. Humpage, K. P. Wong, and Y. W. Lee, "Numerical integration algorithms in power-system dynamic analysis," *Proc. Inst. Electr. Eng. UK*, vol. 121, no. 6, p. 467, 1974, doi: 10.1049/piee.1974.0113.

[408] C. W. Gear, *Numerical initial value problems in ordinary differential equations,* 2nd ed. Englewood Cliffs, NJ: Prentice-Hall, 1971.

[409] E. Hairer, S. P. Nørsett, and G. Wanner, *Solving Ordinary Differential Equations I: Nonstiff problems,* 2nd ed. Berlin: Springer, 2008.

[410] E. Hairer and G. Wanner, *Solving Ordinary Differential Equations II: Stiff and Differential-Algebraic Problems,* 2nd ed. Berlin, Heidelberg: Springer-Verlag Berlin Heidelberg, 2010. [Online]. Available: http://dx.doi.org/10.1007/978-3-642-05221-7

[411] J. Beerten and R. Belmans, "MatACDC - an open source software tool for steady-state analysis and operation of HVDC grids," in *11th IET International Conference on AC and DC Power Transmission*, Birmingham, UK, 2015, 088 (9 .)-088 (9 .).

[412] IEEE Committee Report, "Dynamic Models for Steam and Hydro Turbines in Power System Studies," *IEEE Trans. on Power Apparatus and Syst.*, PAS-92, no. 6, pp. 1904–1915, 1973, doi: 10.1109/TPAS.1973.293570.

[413] *Energy management system application program interface (EMS-API) – Part 302: Common information model (CIM) dynamics*, IEC 61970-302:2018, 2018.

[414] Power System Dynamic Performance Committee Power, System Stability Subcommittee, Task Force on Turbine-Governor Modeling, "Dynamic Models for Turbine-Governors in Power System Studies," IEEE Power & Energy Society PES-TR1, Jan. 2013. Accessed: Sep. 21 2020. [Online]. Available: https://site.ieee.org/fw-pes/files/2013/01/PES_TR1.pdf

[415] *IEEE Recommended Practice for Excitation System Models for Power System Stability Studies*, 2016.

[416] G. R. Berube, L. M. Hajagos, and R. E. Beaulieu, "A utility perspective on under-excitation limiters," *IEEE Trans. Energy Convers.*, vol. 10, no. 3, pp. 532–537, 1995, doi: 10.1109/60.464878.

[417] S. Cole, J. Beerten, and R. Belmans, "Generalized Dynamic VSC MTDC Model for Power System Stability Studies," *IEEE Trans. Power Syst.*, vol. 25, no. 3, pp. 1655–1662, 2010, doi: 10.1109/TPWRS.2010.2040846.

[418] J. Beerten, "MATACDC 1.0 User's Manual," Jul. 2012. [Online]. Available: https://www.esat.kuleuven.be/electa/teaching/matacdc/MatACDCManual

[419] C. Du, *VSC-HVDC for industrial power systems*. Zugl.: Göteborg, Univ., Diss., 2007. Göteborg: Chalmers Univ. of Technology, 2007.

[420] C. Bajracharya and M. Molinas, "Understanding of Tuning Techniques of Converter Controllers for VSC-HVDC," in *Nordic Workshop on Power and Industrial Electronics (NORPIE), Espoo, Finland*, 2008.

A. Appendix

A.1 Determinations and Definitions

Accuracy: Ratio to measure the error between the estimated value and the true value.

Client: A client is an entity that requests a service from a server and that can receive unsolicited messages from a server [396].

Close to real time: Denotes the planning timeframe below one hour from actual real-time.

Contingency, perturbation, disturbance, and fault: These terms are used interchangeably and refer to large changes in the system state, which may cause congestions, system instability or critical system states.

Control centre / Control room: The site from which a transmission or distribution system operator carries out the system operation of the electric transmission or distribution system.

Cognitive system: A cognitive system can independently determine its own state based on the available information and, through its capability to adapt, learn independently how to achieve certain goals. To achieve this, methods from the field of "artificial intelligence" are applied.

Curative measure: An operational measure that is executed immediately after the occurrence of a foreseeable contingency to mitigate the contingencies impact on power system security [95].

Data: Information represented in a format suitable for automated processing [397].

Digital Twin: A continuously adaptive digital model of a physical system, that allows automatic and bidirectional data exchange with the system in the real world. It has the purpose to monitor, control or optimise the system it mirrors. By learning from the available data including simulation and sensor feedback from the real world, it can support decision making under volatile conditions and is able develop capacities for autonomy.

Dynamic Digital Mirror: A power system modelling engine, which can reflect power system operating conditions and is able to execute steady-state case studies and time-domain simulations.

Energy Management System: A computer system comprising a software platform for basic support functions and a set of applications to ensure the functionality of electrical power generation and transmission equipment in order to ensure adequate security of energy supply at minimum costs [398].

First principle: An assumption deduced from physical laws or mathematical rules.

Framework: A framework provides the foundation for independent application development. It is accessible by interfaces and supports development of an modular but integrated software [287].

Frequency stability: "*The ability of a power system to maintain steady frequency following a disturbance resulting in a significant imbalance between generation and load.*" [188, 189]

Latency: The time between when sensor outputs are present at the physical interface of a measuring device until its value is available to the first user or program [133].

Model fidelity: Accuracy of the model in comparison to the actual system state based on measurement data.

***N*-1 criterion:** Redundancy concept in which the power grid must remain secure at all times even if a component fails.

Observability area: The area of responsibility under control by a system operator which is observed by real-time monitoring to maintain operational security.

Observability: Observability is given, if all state variables are determinable for a certain area from a given set of available measurements [399]. A continuous-time system is observable if for any time $t > 0$ the initial state can be uniquely determined by knowledge of the input variables and output variables [185].

Operational technology: The purpose of operational technology is to support industrial process automation.

Power System Stability "*The ability of an electric power system, for a given initial operating condition, to regain a state of operating equilibrium after being subjected to a physical disturbance*" [188, 189].

Preventive measure: An operational measure applied in power system operation and operational planning to reduce the possible number of critical contingencies that cause a violation of operational security limits by optimisation [251].

Process model: „*Reproduction of a process within a simulation-ready model that represents the process adequately enough with reference to selected items.*" [285]

Pseudo measurement: Data generated by a software application, representing the results of a calculation, or states of a process [133].

Real-time processing: Refers to the processing of data by a computer in connection with another process outside the computer according to the time requirements of the external process. This also includes systems, operating in conversational mode, or processes, that can be influenced by human intervention while they are in progress. [336]

Resilience *"The ability to withstand and reduce the magnitude and/or duration of disruptive events, which includes the capability to anticipate, absorb, adapt to, and/or rapidly recover from such an event." [400]*

Rotor angle stability: "*The ability of synchronous machines to remain in synchronism after being subjected to a disturbance.*" [188, 189]

Sandbox: A sandbox is an environment that is isolated from the real system in which actions can be executed and validated. Sandboxes can be used, to test control actions or to protect the underlying system from malicious or harmful actions.

SCADA: A supervisory control and data acquisition system receives telemetry data and thereby allows the system operator to monitor and telecontrol the power system via communication links and standardized protocols.

Server: A server is an entity that communicates with a client and sends information to peer devices [396].

Single source of truth: Refers to an information system design approach, applying information models in a way, that a single information source structures and associates the data elements by referencing, in such way, that data redundancy is avoided.

Situation awareness: "*The perception of the elements in the environment within a volume of time and space, the comprehension of their meaning and the projection of their status in the near future.*" [401]

Virtual Space: A computer-simulated environment that allows interaction via an interface [402].

Voltage stability: "*The ability of a power system to maintain steady voltages at all buses in the system after being subjected to a disturbance.*" [188, 189]

A.2 Coordinate Transformation

A.2.1 αβ0-Transformation

Under certain conditions, coordinate transformations make it easier to represent and calculate mathematical relationships. By the αβ0 transformation, a three-phase power system can be represented in an orthogonal coordinate system by a rotating space vector, also called phasor. This means that the phasors resulting from the instantaneous values V_a, V_b and V_c of three-phase system are expressed by the orthogonal components V_α and V_β and an additional zero component V_0 for balanced systems. In matrix form the transformation is represented as follows [403]:

$$\begin{bmatrix} V_\alpha \\ V_\beta \\ V_0 \end{bmatrix} = \begin{bmatrix} \frac{2}{3} & -\frac{1}{3} & -\frac{1}{3} \\ 0 & \frac{1}{\sqrt{3}} & -\frac{1}{\sqrt{3}} \\ \frac{1}{3} & \frac{1}{3} & \frac{1}{3} \end{bmatrix} \cdot \begin{bmatrix} V_a \\ V_b \\ V_c \end{bmatrix} \tag{A.1}$$

In the symmetrical case, which is always assumed in this work, the zero component is neglected. Thus, only the orthogonal αβ coordinates describe the phasor. This notation has the advantage that arithmetic operations can be carried out faster.

A.2.2 dq-Transformation

In AC power systems, the space vector in αβ coordinates rotates with the angular velocity ω. As a result, the instantaneous values of the α and β components are constantly changing. However, the rotating space vector can also be viewed relative to a coordinate system rotating at the speed ω. In stationary state (ω is equal or close to ω_{ref}), the space vector can be represented, or approximated by constant values in the rotating dq-frame. This dq-transformation (see Park [404]) is usually used to integrate rotating space vectors into a common system frame. Regarding the definitions of the d- and q-axis, different approaches can be found in the literature. The rotor angle δ describes the angular difference between the rotor and the stator field or the complex vector which precedes the stator voltage. It can be defined as the angle between the stator voltage and the d-axis (e.g. [405]) or the q-axis (e.g. [43]) of the rotor. Although it is arbitrary to define the leading rotor axis, this work follows the original description by Park [404], which is also IEEE standard (the quadrature-axis leads the direct-axis [359]). Thus, it is determined that the q-axis leads the d-axis by 90 degrees and the angle δ represents the angle between the real axis of the stator and the d-axis of the rotor. A visual representation of this definition is given in Fig. A.1.

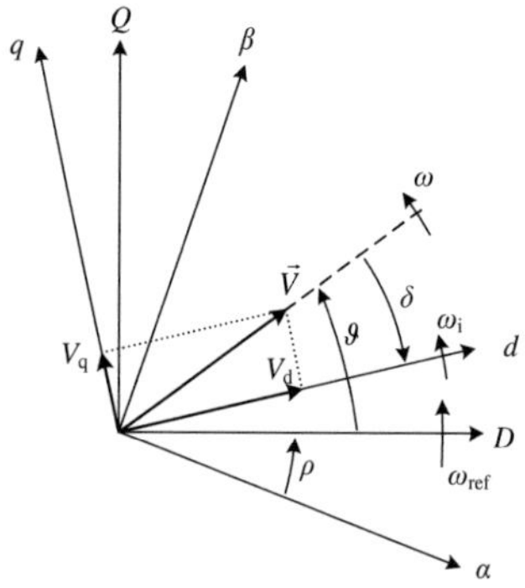

Fig. A.1: Illustration of the quantities defining the reference frame

Note, that the dq-frame is different for each generator and converter in the system. Thus, phasor variables need to be transformed to the d-q domain and transformed back into the system frame, where generators or converters are connected [341].

Different reference coordinate systems are applied within this work. The first coordinate system is the DQ- coordinate system described above and illustrated in Fig. A.1. It is used as reference system in which all voltage space vectors of the AC system are described.

$$\begin{bmatrix} V_D \\ V_Q \end{bmatrix} = \begin{bmatrix} \cos\rho(t) & \sin\rho(t) \\ -\sin\rho(t) & \cos\rho(t) \end{bmatrix} \cdot \begin{bmatrix} V_\alpha \\ V_\beta \end{bmatrix}, \text{ with } \rho(t) = \int \omega_{ref} dt + \rho_0(t) \tag{A.2}$$

The angle $\rho(t)$ represents the instantaneous difference between the D-axis of the rotating coordinate system and the Q-axis of the static reference system.

After the initialization of the dynamic simulation the coordinate system is rotated in such a way that the vector of the node voltage of the reference generator is originated in the D-axis and thus applies to the angle of the voltage vector at the reference generator $\vartheta_{ref}(t_0) = 0$. The reference generator determines the reference angle ρ_0 for the DQ coordinate system. It rotates with the reference frequency ω_{ref} and enables the simplified phasor representation of all AC quantities. All other coordinate systems refer to the DQ coordinate system as reference, i.e., they are described relative to it. The dq coordinate system, which is used within the synchronous machines and inverters (individual reference frame), is related to the DQ coordinate system by the following description according to equation (A.2)

$$\begin{bmatrix} V_d \\ V_q \end{bmatrix} = \begin{bmatrix} \cos(\delta(t)+\vartheta(t)) & \sin(\delta(t)+\vartheta(t)) \\ -\sin(\delta(t)+\vartheta(t)) & \cos(\delta(t)+\vartheta(t)) \end{bmatrix} \cdot \begin{bmatrix} V_D \\ V_Q \end{bmatrix} \tag{A.3}$$

The definition of the angles for the reference frame in VSC-converter models is identical as for the generators.

A.2.3 Numerical Integration Routines

To model the electromechanical dynamics of the power system in the time-domain, a set of differential algebraic equations (DAE) needs to be formulated. These describe the dynamic states of the power system. The dynamic components, such as synchronous machines, converters and dynamic loads are formulated by differential equations (A.4). The algebraic equations (A.5) are required to interface the differential equations to the power system, and describe the network topology including buses, branches, static loads, as well as static generating units.

One distinguishes between explicit and implicit numerical methods for solving differential equation systems. The main difference between both methods is that the explicit methods predict the system states for the following time step from the values of the current states by approximation instead of solving the nonlinear differential equations iteratively. Thereby the computational effort can be reduced, but at the expense of the numerical robustness. In comparison to implicit solution methods, explicit solution methods require smaller simulation time increments, i.e., a higher number of time steps to maintain numerical stability. Their advantage is that the computational effort per time increment is lower in comparison to implicit solution methods since an iterative convergence control is not applied. Thus, within implicit methods, the calculation effort per step is increased by an iterative convergence control, while longer simulation time steps are applicable.

While the advantage of the implicit methods is a higher numerical stability, the main disadvantage is that the entire matrix must be inverted at each time to accommodate new system conditions following events. For large nonlinear systems this can cause a high computation time. Explicit integration methods such as the Modified Euler run significantly faster in comparison to implicit integration methods in HPC implementations for parallel computation [345].

Two general approaches to solve the DAE formulation of power systems can be distinguished [347]. The equations (A.4) and (A.5) can be solved simultaneously in a combined approach or separately in a partitioned solution. The partitioned approach allows to update the differential states $\boldsymbol{x}$ and the algebraic states $\boldsymbol{z}$ separately, but the interface between the differential and algebraic equations may introduce truncation and round-off errors and a computing time delay. It is typically used with explicit numerical methods, which do not require computing and factorisation of the Jacobian matrix [347].

$$\frac{d}{dt}\boldsymbol{x} = \dot{\boldsymbol{x}} = \boldsymbol{f}(\boldsymbol{x}(t), \boldsymbol{z}(t), \boldsymbol{p}) \tag{A.4}$$

$$0 = \boldsymbol{g}(\boldsymbol{x}(t), \boldsymbol{z}(t), \boldsymbol{p}) \tag{A.5}$$

$$\boldsymbol{y}(t) = \boldsymbol{h}(\boldsymbol{x}(t), \boldsymbol{z}(t), \boldsymbol{p}) \tag{A.6}$$

The simultaneous approach jointly updates the differential and algebraic state variables $\boldsymbol{x}$, $\boldsymbol{z}$ by solving the set of nonlinear equations requires an iterative approach to compute and refactorize the Jacobian matrix, which is the reason why the simultaneous approach is slower, but inherits a higher numerical stability in comparison to the partitioned approach [347].

The partitioned approach can be considered faster than the simultaneous approach, but its numerical stability is lower due to the need to interface algebraic and differential equations instead of formulating the power system equations in such way, that allows to solve them in one step [347]. Either way, the solution needs to satisfy the formulated equations, thus the time scale of the dynamics included in the set of DAE needs to be considered in order to select the appropriate method and integration step size [338].

The numerical stability of an integration method depends on the *stiffness* of the given differential equations, which are needed to represent the system [5]. The term *stiffness* is closely related to the computation time required by the approximation method used. "*A problem is stiff if the solution being sought varies slowly, but there are nearby solutions that vary rapidly, so the numerical method must take small steps to obtain satisfactory results*" [406]. In other words, stiffness is an efficiency problem of DAE solvers, because the solution of a stable system becomes unstable when subjected to a certain numerical approximation method [344].

For simulation models in the power system time-domain, the *stiffness* depends on the time constants within the system model and can be characterized by an eigenvalue analysis [5]. In general, the chosen time steps should be smaller than the time constants of the system equipment. For models with a higher detail level (e.g., high order generator models, complex controllers or considering electromagnetic transients), the stiffness of the differential equation increases. Thus, smaller step sizes are required to solve the system by explicit integration methods. This leads to a significantly higher computation time. Due to truncation errors caused by the mathematical approximation of the chosen method and round-off errors caused by digital processing, deviations from the exact solution cannot be avoided during the calculation. These errors can lead to numerical stability issues. The system is called numerically unstable if the error increases from time step to time step. In this case, no information can be drawn about the actual stability of the modelled system.

The integration rules implemented into the DDM simulation engine to solve the power system time-domain models (see subsection 4.3.1) are explained in brief below. A more detailed discussion can be found in [5, 347].

Trapezoidal Rule

With h representing the integration step size and k representing the actual time step, the implicit trapezoidal integration technique formulates as follows:

$$y_{k+1} = y_k \frac{h}{2}(f(y_k) + f(y_{k+1})),\ k \in \mathbb{N} \tag{A.7}$$

The method has advantages in numerically stability (absolute stable), allowing a larger step size. Its drawback lies in its computation time, since y_{k+1} needs to be calculated iteratively.

Euler Method

The explicit single-step Euler method formulates as follows:

$$y_{k+1} = y_k + h\, f(y_k),\ k \in \mathbb{N} \tag{A.8}$$

The approximated results of the single-step Euler method are not very accurate. Thus, it is mostly applied in a modified version.

Modified Euler

The explicit modified Euler method is a two stage predictor and corrector method, which is widely used in power system analysis [5]. To overcome the inaccuracies of the standard Euler method, it averages the derivatives at the two ends of the time increment. This yields a higher approximation accuracy in comparison to the single-step Euler method, which uses only the derivatives at the beginning of the time increment.

$$y_{k+1} = y_k + h\,f(y_k + \frac{h}{2} f(y)),\ k \in \mathbb{N} \tag{A.9}$$

Expressed in terms of a time-domain simulation, the modified Euler consists of the following two steps [5, 347]:

Predictor step:

$$y_{k+1}(h_0) = y_k + h\,f(y_k) \tag{A.10}$$

Corrector Step:

$$y_{k+1}(h_1) = y_k + \frac{h}{2}\left(f(y_k) + f(y_{k+1}(h_0))\right) \tag{A.11}$$

Runge-Kutta

The explicit Runge-Kutta integration method is a multi-stage approach. Although higher order methods can be of advantage for power system simulations in some cases, the four-stage Runge-Kutta variant (A.13)-(A.17) has been proven to be efficient and accurate for application in power systems computation [407]. The 4th order Runge-Kutta method can be represented by a Butcher tableau (A.12), which contains the coefficients c_n, which are chosen to minimise computing effort [344]:

$$\begin{array}{c|cccc} 0 & & & & \\ \frac{1}{2} & \frac{1}{2} & & & \\ \frac{1}{2} & 0 & \frac{1}{2} & & \\ 1 & 0 & 0 & 1 & \\ \hline & \frac{1}{6} & \frac{2}{6} & \frac{2}{6} & \frac{1}{6} \end{array} \tag{A.12}$$

The Runge-Kutta method formulates as follows [347]:

$$c_0 = f(y_k) \tag{A.13}$$

$$c_1 = f(y_k + \frac{1}{2} h\, c_0) \tag{A.14}$$

$$c_2 = f(y_k + \frac{1}{2} h\, c_1) \tag{A.15}$$

$$c_3 = f(y_k + h\, c_2) \tag{A.16}$$

$$y_{k+1} = y_k + \frac{h}{6}(c_0 + 2c_1 + 2c_2 + c_3) \tag{A.17}$$

Summarizing the explanations given beforehand concerning numerical integration methodologies, the main advantage of explicit methods, is their reasonable implementation effort and their performance in terms of computation time [345]. Their main drawback is, that explicit integration methods are not numerically A-stable [5], [408]. Their applicability is therefore limited to non-stiff DAE, or models with moderate stiffness, such as time-domain simulations in the phasor domain [341]. Stiff DAE models may cause numerical problems in terms of solvability due to this limitation. Thus, the chosen step size *h* to solve a set of system state equations is bound to the smallest time constant in the system. Due to the small, fixed step size, long-term simulation scenarios may require long simulation times to run through. To solve stiff systems or long term simulations, routines with variable step size should be chosen, such as Runge-Kutta-Fehlberg [409] or Higham and Hall [410]. These numerical solvers adjust their step size automatically to guarantee a solution accuracy within a certain quality, while maintaining reasonable computation time.

A.3 Time-domain Simulation Models

The *MATLAB* implementation of the dynamic digital mirror (DDM) simulation engine is partly based on available open-source software, respectively *MATPOWER* [369], *MatACDC* [411], and *MatDyn* [346]. Whereas *MATPOWER* and *MatACDC* are applied to solve the power flow equations of the AC and the DC grid respectively. As the power system dynamics are in focus, the simulation environment *MatDyn* has been extended as described within this thesis to accommodate the DDM approach (see also [356]). As the simulation engine can run time-domain simulations of hybrid AC-DC grids offline as well, the standalone simulation engine is called *MatDynACDC* to respect its origins. The time-domain model of the sixth order synchronous machine, the VSC-HVDC transmission system, and the related controllers have been partly described before in [356]. The applied model equations are thoroughly described in the following subsections.

A.3.1 Sixth Order Synchronous Generator Model

The sixth order synchronous machine has been derived from the technical reference of the commercial power systems computation software DIgSILENT PowerFactory, where the model is applied as standard model for studies of electromechanical and electromagnetic transient power system dynamics [385]. As illustrated in Fig. A.2, the equivalent circuit of the synchronous machine d-axis comprises two rotor loops representing the excitation field winding and the 1d-damper winding. The q-axis of a round-rotor machine model is illustrated as well with the 1q- and 2q-damper windings. Both comprise an additional mutual magnetising reactance, which are obtained from the synchronous and leakage reactance [385].

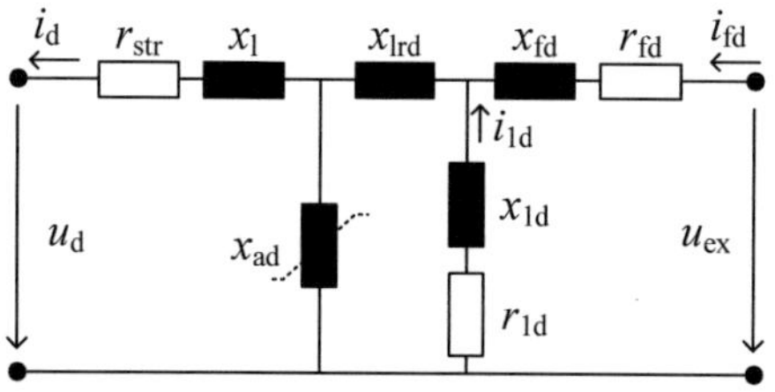

Fig. A.2: Equivalent circuit of the round rotor synchronous machine d-axes and q-axes (adapted from [385])

The system dynamics can be described by the differential equations (A.18) to (A.23),

$$\frac{d}{dt}n(t) = \frac{1}{2H}(t_m(t) - t_e(t) - t_d(t)) \tag{A.18}$$

$$\frac{d}{dt}\delta(t) = \omega_n(n(t) - n_{ref}) \tag{A.19}$$

$$\frac{d}{dt}\psi_{fd}(t) = \omega_n(u_{fd}(t) - r_{fd} \cdot i_{fd}(t)) \tag{A.20}$$

$$\frac{d}{dt}\psi_{1d}(t) = -\omega_n \cdot r_{1d} \cdot i_{1d}(t) \tag{A.21}$$

$$\frac{d}{dt}\psi_{1q}(t) = -\omega_n \cdot r_{1q} \cdot i_{1q}(t) \tag{A.22}$$

$$\frac{d}{dt}\psi_{2q}(t) = -\omega_n \cdot r_{2q} \cdot i_{2q}(t) \tag{A.23}$$

And the algebraic equations (A.24) to (A.37)

$$\psi''_d(t) = k_{fd} \cdot \psi_{fd}(t) + k_{1d} \cdot \psi_{1d}(t) \tag{A.24}$$

$$\psi''_q(t) = k_{1q} \cdot \psi_{1d}(t) + k_{2q} \cdot \psi_{2q}(t) \tag{A.25}$$

$$\psi_d(t) = -x''_d \cdot i_d(t) + \psi''_d(t) \tag{A.26}$$

$$\psi_{q}(t)=-x''_{q}\cdot i_{q}(t)+\psi''_{q}(t) \tag{A.27}$$

$$t_{m}(t)=\frac{p_{t}(t)}{n(t)}-d\cdot n(t) \tag{A.28}$$

$$t_{e}(t)=\frac{1}{\cos(\varphi_{n})}(i_{q}(t)\cdot\psi_{d}(t)-i_{d}(t)\cdot\psi_{q}(t)) \tag{A.29}$$

$$t_{d}(t)=d(n(t)-n_{ref}) \tag{A.30}$$

$$i_{1d}(t)=k_{1d}\cdot i_{d}(t)+\frac{x_{fd,loop}}{x_{det,d}}\cdot\psi_{1d}(t)-\frac{x_{ad}+x_{rld}}{x_{det,d}}\cdot\psi_{fd}(t) \tag{A.31}$$

$$i_{fd}(t)=k_{fd}\cdot i_{d}(t)+\frac{x_{1d,loop}}{x_{det,d}}\cdot\psi_{fd}(t)-\frac{x_{ad}+x_{rld}}{x_{det,d}}\cdot\psi_{1d}(t) \tag{A.32}$$

$$i_{1q}(t)=k_{1q}\cdot i_{q}(t)+\frac{x_{2q,loop}}{x_{det,q}}\cdot\psi_{1q}(t)-\frac{x_{aq}+x_{rlq}}{x_{det,q}}\cdot\psi_{2q}(t) \tag{A.33}$$

$$i_{2q}(t)=k_{2q}\cdot i_{q}(t)+\frac{x_{1q,loop}}{x_{det,q}}\cdot\psi_{2q}(t)-\frac{x_{aq}+x_{rlq}}{x_{det,q}}\cdot\psi_{1q}(t) \tag{A.34}$$

$$u_{d}(t)=-n(t)\cdot\psi''_{q}(t)-r_{str}\cdot i_{d}(t)+\hat{n}(t)\cdot x''_{q}\cdot i_{q}(t) \tag{A.35}$$

$$u_{q}(t)=n(t)\cdot\psi''_{d}(t)-r_{str}\cdot i_{q}(t)-\hat{n}(t)\cdot x''_{d}\cdot i_{d}(t) \tag{A.36}$$

$$u_{fd}(t)=\frac{r_{fd}}{x_{ad}}\cdot v_{ex}(t) \tag{A.37}$$

The speed reference value $n_{ref}=1$ p.u., and $\hat{n}(t)=1$, since the model in the phasor domain partly neglects the rotor speed dependency (see also [5]). The transformation into the synchronous d-q frame allows to consider the rotating d-q components as constant, assuming, that they rotate with the reference frequency ω_{ref}. Following the model description in [385], the q-axis leads the d-axis by 90 degrees. Thus, the rotor angle of the machine is referenced to the terminal voltage angle θ, and the state variable δ is initialised as: $\delta = \arctan(\underline{u}_t + (r_{str} + jx_q)\cdot\underline{i}_t) - \pi/2$. Where $\underline{u}_t$ is the generator terminal voltage, r_{str} is the stator resistance, x_q is the synchronous reactance of the machine's q-axis, and $\underline{i}_t$ is the current injected into the grid. To calculate the voltage at the generator terminal, the subtransient stator admittance is applied within the augmented admittance matrix $\underline{\boldsymbol{Y}}_{aug}$. By neglecting subtransient saliency (i.e., $x'' = x''_q = x''_d$), the stator admittance $\underline{\boldsymbol{Y}}_g$ can be expressed as $\underline{\boldsymbol{Y}}_g = (r_{str} + jx'')^{-1}$. To consider a saliency within the generator admittance $\underline{\boldsymbol{Y}}_g$, a dummy rotor coil can be assumed, which is represented by an additional state variable [358]. The complex current injection by the generator at the generator terminal into the grid can be calculated by (A.38) as follows:

$$i_{\mathrm{d}}(t)+ji_{\mathrm{q}}(t)\cdot e^{j\delta(t)}=\underline{\mathbf{Y}}_{\mathrm{g}}(u''_{\mathrm{d}}(t)-u_{\mathrm{d}}(t)+j(u''_{\mathrm{q}}(t)-u_{\mathrm{q}}(t)))\cdot e^{j\delta(t)} \tag{A.38}$$

Stacking the difference equations describing the differential states $x=(n,\ \delta,\ \psi_{\mathrm{fd}},\ \psi_{\mathrm{1d}},\ \psi_{\mathrm{1q}},\ \psi_{\mathrm{2q}},)^{\mathrm{T}}$: $[0,\infty)\rightarrow\mathbb{R}^6$, the algebraic states $z=(i_{\mathrm{d}},\ i_{\mathrm{q}})^{\mathrm{T}}:[0,\infty)\rightarrow\mathbb{R}^2$, and the inputs $u=(p_{\mathrm{t}},\ u_{\mathrm{ex}})^{\mathrm{T}}$: $[0,\infty)\rightarrow\mathbb{R}^2$ given in (A.18) - (A.38) into a DAE formulation, these can be summarised as follows:

$$\begin{aligned}\frac{d}{dt}x_1(t)=&\frac{1}{2H}\left(\frac{u_1(t)}{x_1(t)}-d\cdot x_1(t)-d\cdot(x_1(t)-n_{\mathrm{ref}}\right)\\&-\frac{1}{\cos(\varphi_{\mathrm{n}})}[z_2(t)\cdot(-x''_{\mathrm{d}}\cdot z_1(t)+k_{\mathrm{fd}}\cdot x_3(t)+k_{\mathrm{1d}}\cdot x_4(t))\\&\qquad-z_1(t)\cdot(-x''_{\mathrm{q}}\cdot z_2(t)+k_{\mathrm{1q}}\cdot x_5(t)+k_{\mathrm{2q}}\cdot x_6(t))])\end{aligned} \tag{A.39}$$

$$\frac{d}{dt}x_2(t)=\omega_{\mathrm{n}}\cdot(x_1(t)-n_{\mathrm{ref}}) \tag{A.40}$$

$$\frac{d}{dt}x_3(t)=-\omega_{\mathrm{n}}\left[r_{\mathrm{fd}}\cdot\left(k_{\mathrm{fd}}\cdot z_1(t)+\frac{x_{\mathrm{1d,loop}}}{x_{\mathrm{det,d}}}\cdot x_3(t)-\frac{x_{\mathrm{ad}}+x_{\mathrm{rld}}}{x_{\mathrm{det,d}}}\cdot x_4(t)\right)-\frac{r_{\mathrm{fd}}}{x_{\mathrm{ad}}}\cdot u_2(t)\right] \tag{A.41}$$

$$\frac{d}{dt}x_4(t)=-\omega_{\mathrm{n}}\cdot -r_{\mathrm{1d}}\cdot\left(k_{\mathrm{1d}}\cdot z_1(t)+\frac{x_{\mathrm{fd,loop}}}{x_{\mathrm{det,d}}}\cdot x_4(t)-\frac{x_{\mathrm{ad}}+x_{\mathrm{rld}}}{x_{\mathrm{det,d}}}\cdot x_3(t)\right) \tag{A.42}$$

$$\frac{d}{dt}x_5(t)=-\omega_{\mathrm{n}}\cdot -r_{\mathrm{1q}}\cdot\left(k_{\mathrm{1q}}\cdot z_2(t)+\frac{x_{\mathrm{2q,loop}}}{x_{\mathrm{det,q}}}\cdot x_5(t)-\frac{x_{\mathrm{aq}}+x_{\mathrm{rlq}}}{x_{\mathrm{det,q}}}\cdot x_6(t)\right) \tag{A.43}$$

$$\frac{d}{dt}x_6(t)=-\omega_{\mathrm{n}}\cdot -r_{\mathrm{2q}}\cdot\left(k_{\mathrm{2q}}\cdot z_2(t)+\frac{x_{\mathrm{1q,loop}}}{x_{\mathrm{det,q}}}\cdot x_6(t)-\frac{x_{\mathrm{aq}}+x_{\mathrm{rlq}}}{x_{\mathrm{det,q}}}\cdot x_5(t)\right) \tag{A.44}$$

and algebraic equations

$$\begin{pmatrix}z_1(t)\\z_2(t)\end{pmatrix}=\begin{bmatrix}-r_{\mathrm{str}} & x_1(t)x''_{\mathrm{q}}\\-x_1(t)x''_{\mathrm{d}} & -r_{\mathrm{str}}\end{bmatrix}^{-1}\begin{pmatrix}u_{\mathrm{d}}(t)-u''_{\mathrm{d}}(t)\\u_{\mathrm{q}}(t)-u''_{\mathrm{q}}(t)\end{pmatrix} \tag{A.45}$$

With u_d and u_q given by equation (A.35)-(A.36) and (A.46)-(A.47).

$$u''_{d}(t) = -x_1(t)\left(k_{1q}x_5(t) + k_{2q}x_6(t)\right) \tag{A.46}$$

$$u''_{q}(t) = x_1(t)\left(k_{fd}x_3(t) + k_{1d}x_4(t)\right) \tag{A.47}$$

The corresponding outputs $y = (n, \delta, i_{ex}, \underline{i}_t, \underline{u}_t) : [0, \infty) \rightarrow \mathbb{R}^3, \mathbb{C}^2$ are given by:

$$y_1(t) = x_1(t) \tag{A.48}$$

$$y_2(t) = x_2(t) \tag{A.49}$$

$$y_3(t) = x_{ad}\left(k_{fd} \cdot z_1(t) + \frac{x_{1d,loop}}{x_{det,d}} \cdot x_3(t) - \frac{x_{ad} + x_{rld}}{x_{det,d}} \cdot x_4(t)\right) \tag{A.50}$$

$$y_4(t) = \left(z_1(t) + jz_2(t)\right)^{jx_2(t)} \tag{A.51}$$

$$y_5(t) = \left(u_d(t) + ju_q(t)\right)^{jx_2(t)} \tag{A.52}$$

A.3.1.1 Turbine Speed-Governor Model

Most synchronous machines in the electric power supply system are propelled by turbines, which are either steam-, gas-, or water-driven. These are connected to the synchronous generator via a common shaft to convert mechanical energy into electrical energy. Each turbine requires a controller, for starting and controlled ramping of the turbine power. During normal operation, the controller ensures that the power setpoint and the generator speed are maintained synchronously with the system frequency by adjusting the mechanical power output. A simple model which can represent mechanical-hydraulic and electrical-hydraulic turbine controllers is shown in Fig. A.3 [412], [413]. The model is applicable for mechanical-hydraulic and electric-hydraulic turbine controllers with or without steam recirculation [413]. It comprises signal inputs for primary and secondary control. Guidelines for the application of dynamic models for turbine-governors in power system studies are given in [414]. The block diagram illustrated in Fig. A.3 has been translated into its subordinate differential equations for straight forward implementation into the modelling engine. The parameter set of the controller model is documented in the appendix A.5.

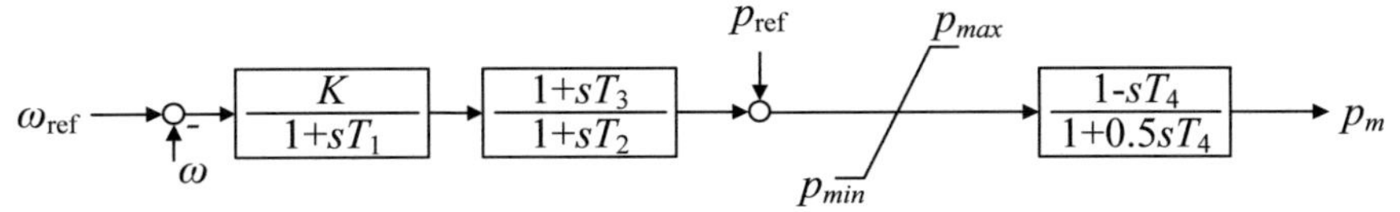

Fig. A.3: Illustration of a simple speed governor for synchronous machines [412], [413]

The mechanical turbine power output can be computed as follows:

$$\frac{d}{dt}x_{1,\mathrm{gov}}(t)=\frac{1}{T_{1,\mathrm{gov}}}\left(K_{\mathrm{gov}}\cdot\Delta n(t)-x_{1,\mathrm{gov}}(t)\right) \tag{A.53}$$

$$\frac{d}{dt}x_{2,\mathrm{gov}}(t)=\frac{1}{T_{3,\mathrm{gov}}}\left(x_{1,\mathrm{gov}}(t)-x_{2,\mathrm{gov}}(t)\right) \tag{A.54}$$

$$\frac{d}{dt}x_{3,\mathrm{gov}}(t)=\frac{2}{T_{4,\mathrm{gov}}}\left[\prod\left(x_{2,\mathrm{gov}}(t)+\frac{T_{2,\mathrm{gov}}}{T_{3,\mathrm{gov}}}\left(x_{1,\mathrm{gov}}(t)-x_{2,\mathrm{gov}}(t)\right)+u_{\mathrm{gov}}(t)\right)-x_{3,\mathrm{gov}}(t)\right] \tag{A.55}$$

$$p_{\mathrm{t}}(t)=3\cdot x_{3,\mathrm{gov}}(t)-2\cdot\left[\prod\left(x_{2,\mathrm{gov}}(t)+\frac{T_{2,\mathrm{gov}}}{T_{3,\mathrm{gov}}}\left(x_{1,\mathrm{gov}}(t)-x_{2,\mathrm{gov}}(t)\right)+u_{\mathrm{gov}}(t)\right)\right] \tag{A.56}$$

Where $\prod$: $\mathbb{R}\rightarrow[p_{\min,\mathrm{gov}}, p_{\max,\mathrm{gov}}]$, $p\rightarrow(\min(\max(p, p_{\min,\mathrm{gov}}), p_{\max,\mathrm{gov}})$ denotes the projection to the interval $[p_{\min,\mathrm{gov}}, p_{\max,\mathrm{gov}}]$ and $u_{\mathrm{gov}} = p_{\mathrm{gov,ref}}$.

A.3.1.2 Automatic Voltage Regulator (AVR) Model

The primary voltage of the synchronous machine is regulated by the automatic voltage regulator (AVR). It regulates of the excitation field voltage v_{ex}. A large number of standardized excitation system models for power system stability studies exists. A good overview about excitation systems and their application is provided by the IEEE Standard 421.5 [298]. Fig. A.4 illustrates the excitation system IEEE AC4C. It has been implemented into the power system modelling engine and enables the integration of stator current limiters as well as over- and under-excitation limiters. The parameters of the excitation system are documented in the appendix A.5.

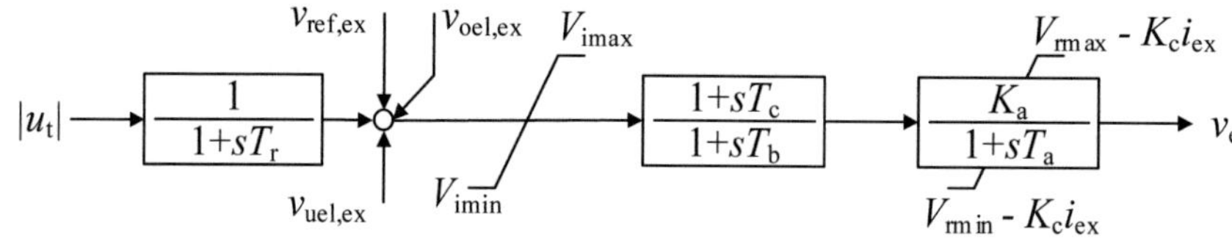

Fig. A.4: Signal flow diagram of the excitation system IEEE AC4C [415]

The equation system of the IEEE AC4C-Model is given below. It takes the generators excitation current i_{ex} as reference to regulate the excitation voltage v_{ex}. The differential-algebraic system of equations for excitation control can be formulated as:

$$\frac{d}{dt}x_{1,ex}(t) = \frac{1}{T_{r,ex}}\left(|\underline{u}_t(t)| - x_{1,ex}(t)\right) \tag{A.57}$$

$$\frac{d}{dt}x_{2,ex}(t) = \frac{1}{T_{b,ex}}\left[\prod\left(v_{ref} - x_{1,ex}(t) + v_{uel} + v_{oel}\right)\right] - x_{2,ex}(t) \tag{A.58}$$

$$\frac{d}{dt}x_{3,ex}(t) = \frac{1}{T_{a,ex}}\left(K_a \cdot x_{2,ex}(t) + \left(\frac{T_c}{T_b} \cdot \left[\prod\left(v_{ref} - x_{1,ex}(t) + v_{uel} + v_{oel}\right)\right] - x_{2,ex}(t)\right) - x_{3,ex}(t)\right) \tag{A.59}$$

$$v_{ex}(t) = \max\left(\min\left(x_{3,ex}(t), V_{r,max} - K_c \cdot i_{ex}\right), V_{r,min} - K_c \cdot i_{ex}\right) \tag{A.60}$$

Where $\prod$: $\mathbb{R} \rightarrow [V_{i,min,ex}, V_{i,max,ex}]$, $V \rightarrow (\max(\min(V, V_{i,max,ex}), V_{i,min,ex})$ denotes the projection to the interval $[V_{i,min,ex}, V_{i,max,ex}]$. Note that the external signals of the overexcitation limiter $v_{oel,ex}$, and the under-excitation limiter $v_{uel,ex}$ are considered as zero in case there is no signal connected. The external signal $v_{ref,ex}$ represents the voltage setpoint at the generator bus and $|u_t|$ the voltage magnitude measured at the generator terminal.

Excitation Limiter

To represent the operating limits of the synchronous machine in the time-domain simulation correctly, it is necessary to monitor over- and under-excitation by limiting controllers. The function of these is briefly described below.

Over-excitation Limiter

The balancing of voltage sags in dynamic simulations of transmission grids requires the modelling of an overexcitation limiter. The limiters control the excitation at high machine load and are therefore also called maximum excitation limiters, and field current limiters [415]. In this work, the type OverexcLim2 is used, which is shown in Fig. A.5. It has a fixed threshold value and reduces the excitation setpoint value by means of an anti-windup integral controller [413].

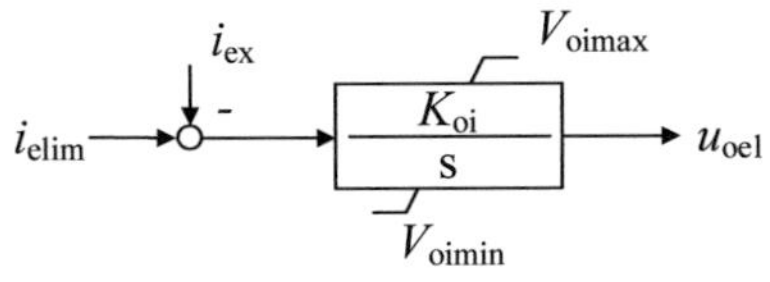

Fig. A.5: Signal flow diagram of the over-excitation limiting system OverExLim2 (adapted from [413])

Under-excitation Limiter

The purpose of the under-excitation limiter is to keep the excitation current above a minimum value, in order to prevent the synchronous machine from becoming unstable or losing synchronisation with the grid as a result of insufficient excitation [415], [416]. Operating states which can lead to thermal overload of the stator or the triggering of the "loss-of-excitation" relay during a transient operating state in the under-excited range are also avoided by the under-excitation limiter [416]. The type OverexcLim2 applied in this work is shown in Fig. A.6.

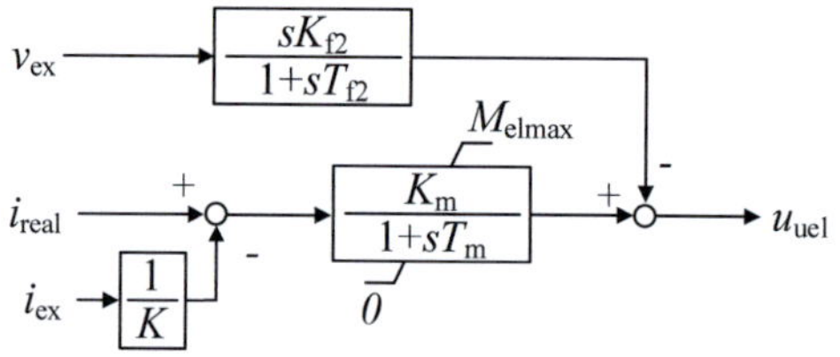

Fig. A.6: Signal flow diagram of the under-excitation limiting system UnderExLimX1 (adapted from [413])

Power system stabilizers (PSS) are not applied in this thesis, since the damping of sub-synchronous resonances is not in scope of the work. For model extensions considering PSS models, the IEEE standard 421.5 [415], is a good reference. Model parameters for synchronous machines are well documented in [299], and in [298] for their excitation systems respectively. Most standardised models are also referenced in the CIM dynamic model profile IEC 61970-302 [413].

A.3.2 Voltage Source Converter High Voltage DC model

The VSC-HVDC dynamics on the AC-side of the converter are modelled according to the equivalent circuit shown in Fig. A.7 (see also Fig. 5.2 in subsection 5.1.2). The VSC-HVDC model is implemented according to the description given in Chaudhuri [365] and Cole [341].

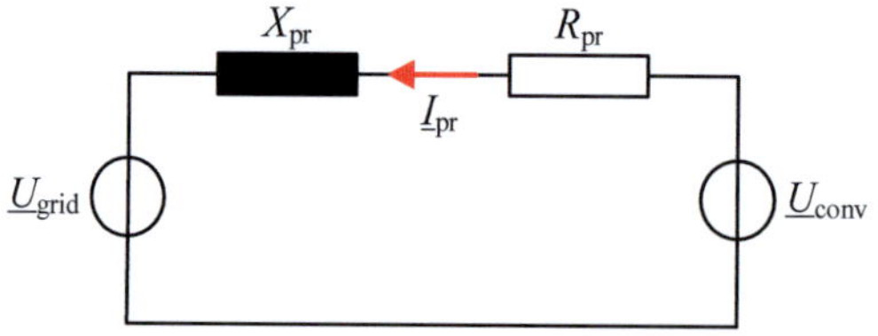

Fig. A.7: Single line diagram of the VSC AC circuit (adapted from [417]).

Following the model equations and descriptions in [417], [341], [365], the VSC-HVDC converters of the generalised model are connected to the power system by a transformer and a phase reactor as illustrated in Fig. A.7. The circuit equation (A.61) can be transformed into dq-components of the rotating reference frame as described in A.2.2.

$$u_{\mathrm{conv}}(t)-u_{\mathrm{grid}}(t)=L_{\mathrm{pr}}\frac{d}{dt}i_{\mathrm{pr}}(t)+R_{\mathrm{pr}}i_{\mathrm{pr}}(t) \tag{A.61}$$

$$\frac{d}{dt}i_{\mathrm{d,pr}}(t)=-\frac{R_{\mathrm{pr}}}{L_{\mathrm{pr}}}\cdot i_{\mathrm{d,pr}}(t)+\omega\cdot i_{\mathrm{q,pr}}(t)+\frac{1}{L_{\mathrm{pr}}}(u_{\mathrm{d,conv}}(t)-u_{\mathrm{d,grid}}(t)) \tag{A.62}$$

$$\frac{d}{dt}i_{\mathrm{q,pr}}(t)=-\frac{R_{\mathrm{pr}}}{L_{\mathrm{pr}}}\cdot i_{\mathrm{q,pr}}(t)-\omega\cdot i_{\mathrm{d,pr}}(t)+\frac{1}{L_{\mathrm{pr}}}(u_{\mathrm{q,conv}}(t)-u_{\mathrm{q,grid}}(t)) \tag{A.63}$$

The coupling between the AC and DC side of the converter is realised by considering the power balance. Thus, the power fed into the AC grid must be equal to DC side including the losses of the converter. The converter losses can be considered by the loss factors a, b and c according to [418].

$$P_{\mathrm{AC}}(t)=u_{\mathrm{d}}(t)\cdot i_{\mathrm{d}}(t)+u_{\mathrm{q}}(t)\cdot i_{\mathrm{q}}(t) \tag{A.64}$$

$$Q_{\mathrm{AC}}(t)=u_{\mathrm{q}}(t)\cdot i_{\mathrm{d}}(t)+u_{\mathrm{d}}(t)\cdot i_{\mathrm{q}}(t) \tag{A.65}$$

$$P_{\mathrm{DC}}=P_{AC}+P_{Loss} \tag{A.66}$$

$$P_{\mathrm{Loss}}=a+b\cdot i_{\mathrm{AC,VSC}}+c\cdot i^{2}{}_{\mathrm{AC,VSC}} \tag{A.67}$$

The DC voltage u_{DC} is determined by the capacitors of the converter circuit C_{DC} and the DC currents. Imbalances between the AC and the DC power cause voltage changes in the DC link capacitor. The DC grid dynamics are given by the following expression:

$$\frac{d}{dt}u_{\mathrm{DC}}(t)=\frac{1}{C_{\mathrm{DC}}}\left(i_{\mathrm{DC,conv}}(t)-i_{\mathrm{DC,grid}}(t)\right) \tag{A.68}$$

$$L_{\mathrm{DC},ij}\cdot\frac{d}{dt}i_{\mathrm{VSC},ij}+R_{\mathrm{DC},ij}\cdot i_{\mathrm{VSC},ij}=u_{\mathrm{DC},i}-u_{\mathrm{DC},j},\text{where } i<j \tag{A.69}$$

Assuming that the *Phase-Locked Loop* (PLL) is fast enough to track the system frequency, it can be neglected for stability studies in the electromechanical time-domain [363], [365]. It is assumed to be ideal and therefore not explicitly modelled.

A.3.2.1 Inner Control Loop (Current Control)

To obtain a modular VSC model as described in subsection 5.1.2, the dynamic equations of the converter and the controllers have been decoupled. For the applied converter model, a proportional-integral (PI) controller is suitable for the inner control loop. The dynamics of the PI controller can be described as follows [365]:

$$\frac{d}{dt}\kappa_{\mathrm{d,VSC}} = i_{\mathrm{dref,VSC}} - i_{\mathrm{d,VSC}} \tag{A.70}$$

$$\frac{d}{dt}\kappa_{\mathrm{q,VSC}} = i_{\mathrm{qref,VSC}} - i_{\mathrm{q,VSC}} \tag{A.71}$$

By compensating the terms that couple the d- and q- components of the rotating reference frame, the controller allows independent control of the converter variables $i_{\mathrm{d,VSC}}$ and $i_{\mathrm{q,VSC}}$. According to [341, 365], the decoupled inner control loop results accordingly in:

$$u_{\mathrm{d,vsc}} = K_{\mathrm{p,VSC}}(i_{\mathrm{dref,VSC}} - i_{\mathrm{d,VSC}}) + K_{\mathrm{i}} \cdot \kappa_{\mathrm{d}} + u_{\mathrm{d,VSC}} - \omega L_{\mathrm{pr}} i_{\mathrm{q,VSC}} \tag{A.72}$$

$$u_{\mathrm{q,vsc}} = K_{\mathrm{p,VSC}}(i_{\mathrm{qref,VSC}} - i_{\mathrm{q,VSC}}) + K_{\mathrm{i}} \cdot \kappa_{\mathrm{q}} + u_{\mathrm{q,VSC}} + \omega L_{\mathrm{pr}} i_{\mathrm{d,VSC}} \tag{A.73}$$

Were i_{d}, i_{q} and u_{d}, u_{q} as well as ω are measurands applied within the inner control loop of the converter, and $K_{\mathrm{p,VSC}}$ and $K_{\mathrm{i,VSC}}$ represent the proportional and the integral gain. To implement the principle of modularity consistently, the system of equations for the inner control loop are decoupled. Thus, the converter model design is also compatible with other designs of the inner control loop if the principle of decoupling is also applied.

$$u_{\mathrm{d,VSC}} = K_{\mathrm{p,VSC}}(i_{\mathrm{dref,VSC}} - i_{\mathrm{d,VSC}}) + K_{\mathrm{i}} \cdot \kappa_{\mathrm{d}} \tag{A.74}$$

$$u_{\mathrm{q,VSC}} = K_{\mathrm{p,VSC}}(i_{\mathrm{qref,VSC}} - i_{\mathrm{q,VSC}}) + K_{\mathrm{i}} \cdot \kappa_{\mathrm{q}} \tag{A.75}$$

$$\frac{d}{dt} i_{\mathrm{d,VSC}} = -\frac{R_{\mathrm{pr}}}{L_{\mathrm{pr}}} i_{\mathrm{d,VSC}} + \frac{u_{\mathrm{d,VSC}}}{L_{\mathrm{pr}}} \tag{A.76}$$

$$\frac{d}{dt} i_{\mathrm{q,VSC}} = -\frac{R_{\mathrm{pr}}}{L_{\mathrm{pr}}} i_{\mathrm{q,VSC}} + \frac{u_{\mathrm{q,VSC}}}{L_{\mathrm{pr}}} \tag{A.77}$$

A.3.2.2 Outer Control Loop

The outer control loop of the VSC model implements higher-level control concepts and provides the reference signals $i_{\mathrm{dref,VSC}}$ and $i_{\mathrm{qref,VSC}}$ for the inner control loop. The decoupling of the d- and q-axes within the inner control loop, allows separate control of the active and reactive power flow into the AC grid, i.e., the power transfer of the VSC-HVDC system. It enables different control concepts for adjusting the active power flow $P_{\mathrm{VSC,AC}}$ at the AC terminal, the power flow $P_{\mathrm{VSC,DC}}$ within the DC grid, or to control the DC voltage u_{dc}.

Furthermore, and independent control of the reactive power flow Q_{VSC} or the regulation of the AC voltage $\underline{u}_{VSC}$ is possible. The VSC-HVDC outer control loop and possible implementations are described in depth in [341], [365] and [419]. A limitation of the active and reactive power according to a voltage dependent capability curve of the VSC is not considered.

A.3.2.3 Active Power Control

To control the active power flow, the reference active power is given by $P_{ACref,VSC}$. The controller characteristic is given by the proportional gain K_p and the integral gain K_i.

$$\frac{d}{dt}\kappa_d = P_{ACref,VSC} - i_{d,VSC} \cdot u_{d,VSC} \tag{A.78}$$

$$i_{dref,VSC} = K_{p,P}(P_{ACref,VSC} - i_{d,VSC} \cdot u_{d,VSC}) + K_{i,P} \cdot \kappa_d \tag{A.79}$$

A.3.2.4 Reactive Power Control

A simple open loop controller is applied to control the reactive power output at the AC terminal of the VSC. While the reference signal $i_{qref,VSC}$ is calculated by the outer control loop, the inner control loop ensures that the reference value of the open loop controller is maintained.

$$i_{qref,VSC} = -\frac{Q_{ref,VSC}}{u_{d,VSC}} \tag{A.80}$$

A.3.2.5 AC-Voltage Control

To maintain a specified voltage at the VSC AC terminal, the reactive power setpoint can be dynamically adapted by specifying a reference voltage $u_{dref,VSC}$. Thereby, $K_{p,\,uac}$ and $K_{i,\,uac}$ are the proportional and integral gain of the PI controller.

$$\frac{d}{dt}\kappa_q = u_{d,VSC} - u_{dref,VSC} \tag{A.81}$$

$$i_{qref,VSC} = K_{p,u_{ac}} \cdot (u_{d,VSC} - u_{ref,VSC}) + K_{i,u_{ac}} \cdot \kappa_q \tag{A.82}$$

A.3.2.6 DC-Voltage control

To control the DC voltage of the VSC, a DC reference voltage $u_{dc,ref}$ can be specified in addition to the proportional and integral gains $K_{p,\,dc}$ and $K_{i,\,dc}$.

$$\frac{d}{dt}\kappa_{\mathrm{d}} = u_{\mathrm{dc}} - u_{\mathrm{dc,ref}} \tag{A.83}$$

$$i_{\mathrm{dref,VSC}} = K_{\mathrm{p,u_{dc}}}(u_{\mathrm{dc}} - u_{\mathrm{dc,ref}}) + K_{\mathrm{i,dc}} \cdot \kappa_{\mathrm{d}} \tag{A.84}$$

A.4 Total Vector Error

The total vector error (TVE) is a measure to describe the error between the theoretical phasor value of the measured input signal value and the phasor estimate [211]. It is applied to evaluate the accuracy of a PMU. The synchrophasor standard IEEE C37.118 [211, 212] specifies, that the TVE of 1 % may not be exceeded for stationary signals. The TVE is calculated according to equation (A.85)(A.1) [211]. Quantities that influence the TVE are the amplitude error, the phase angle error, and the accuracy of the time synchronisation (deviation from UTC). The latter is influenced by the GPS clock and the propagation delay within the GPS receiver. It should be noted that these quantities, i.e., the amplitude error, the phase angle error, and the time delay are indirectly influencing the DDM model accuracy, and might stay undetected, when PMU signals of low quality are applied for model tuning. The TVE determination from the measured signal is illustrated in Fig. A.8.

$$TVE = \sqrt{\frac{(\hat{x}_r(t) - x_r(t))^2 + (\hat{x}_i(t) - x_i(t))^2}{(x_r(t))^2 + (x_i(t))^2}} \tag{A.85}$$

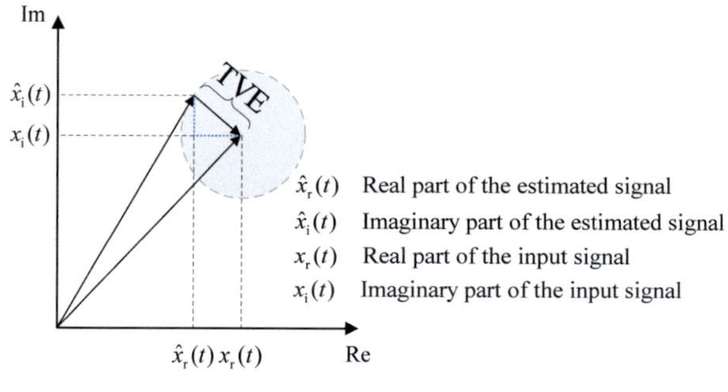

Fig. A.8: Determination of the TVE for PMU measurements according to [211]

A.5 Benchmark Model

The benchmark model is a modified version of the Cigré test system described in [382]. It has been developed to study the influence of embedded HVDC transmission systems on AC system performance and system security [382]. The model design therefore is especially useful to investigate dynamic phenomena, but is partly lacking realistic assumptions (e.g., long transmission lines and large shunts for reactive power compensation). The power system model has been slightly modified, and contains seven power plants, 24 busbars and 36 branches (including transformers). The system can be divided into the two zones "North" and "South", which are interconnected by five AC transmission lines and one HVDC link. The base power flow scenario the northern zone is assigned to have a surplus of generation power, leading to a transit to the southern zone. The base case parameters of the model are documented in Tab. A.1 to Tab. A.14.

Tab. A.1: Bus data of the Cigré benchmark system

Number	Name	U_n kV	Bus Type	Load *P* MW	Load *Q* MVAr	Gen *P* MW	Gen *Q* MVAr
1	NA	400	PQ	-	-	-	-
2	NB	400	PQ	-	-	-	-
3	NC	400	PQ	-	-	-	-
4	ND	400	PQ	500	310	-	-
5	NE	400	PQ	500	310	-	-
6	NF	400	PQ	-	-	-	-
7	NG	400	PQ	1000	620	-	-
8	NH	400	PQ	600	372	-	-
9	SA	400	PQ	1500	930	-	-
10	SB	400	PQ	500	310	-	-
11	SC	400	PQ	500	310	-	-
12	SD	400	PQ	1000	620	-	-
13	SE	400	PQ	-	-	-	-
14	SF	400	PQ	1100	682	-	-
15	SG	400	PQ	-	-	-	-
16	SH	400	PQ	450	280	-	-
17	SI	400	PQ	1250	775	-	-
18	GNA	17	SL	-	-	-	-
19	GNB	17	PV	-	-	1500	-
20	GNF	17	PV	-	-	1120	-
21	GSB	17	PV	-	-	1400	-
22	GSC	17	PV	-	-	870	-
23	GSE	17	PV	-	-	1100	-
24	GSI	17	PV	-	-	1500	-

Tab. A.2: Transmission line type parameters (f_n = 50 Hz)

U_n kV	I_n kA	r' Ω/km	x' Ω/km	l' mH/km	b' µS/km	c' µF/km
400	2,7	0.0128	0.2352	0.7487	4.335	0.0138

Tab. A.3: Reactive power compensation

Name	Bus	Type	Maximum tap position	Active tap position	Q_{max} MVAr	Q_{akt} MVAr
Capacitor NC	NC	Capacitor	10	0	100	0
Capacitor NH	NH	Capacitor	10	2	500	100
Capacitor SF	SF	Capacitor	4	3	2000	1500
Reactor NC	NC	Shunt Reactor	20	0	400	0
Reactor NH	NH	Shunt Reactor	30	0	600	0

Tab. A.4: Transmission lines

No.	Name	Bus *i*	Bus *j*	Length km
1	NA-NB	NA	NB	100
2	NA-NH 1	NA	NH	400
3	NA-NH 2	NA	NH	400
4	NB-NC	NB	NC	50
5	NB-ND 1	NB	ND	50
6	NB-ND 2	NB	ND	50
7	NB-NF	NB	NF	50
8	NC-NE	NC	NE	100
9	NC-SC	NC	SC	800
10	ND-NE 1	ND	NE	20
11	ND-NE 2	ND	NE	20
12	NE-SC	NE	SC	1000
13	NF-NG 1	NF	NG	100
14	NF-NG 2	NF	NG	100
15	NF-SI	NF	SI	800

No.	Name	Bus *i*	Bus *j*	Length km
16	NG-SG	NG	SG	500
17	NH-SA 1	NH	SA	600
18	NH-SA 2	NH	SA	600
19	SA-SB	SA	SB	30
20	SA-SF 1	SA	SF	100
21	SA-SF 2	SA	SF	100
22	SC-SD 1	SC	SD	20
23	SC-SD 2	SC	SD	20
24	SC-SE	SC	SE	20
25	SD-SF	SD	SF	20
26	SE-SF	SE	SF	20
27	SE-SH	SE	SH	35
28	SF-SI	SF	SI	300
29	SG-SH	SG	SH	5

Tab. A.5: Transformers

Name	Bus i	Bus j	S_n MVA	U_n HV	U_n LV	u_k %	u_{kr} %
T_GNA	NA	GNA	2625	400	17	10	0,2
T_GNB	NB	GNB	2250	400	17	10	0,2
T_GNF	NF	GNF	1500	400	17	10	0,2
T_GSB	SB	GSB	1875	400	17	10	0,2
T_GSC	SC	GSC	1125	400	17	10	0,2
T_GSE	SE	GSE	1500	400	17	10	0,2
T_GSI	SI	GSI	2250	400	17	10	0,2

Tab. A.6: Power plant data

No.	Name	Type (see Tab. A.7)	S_n MVA	U_n kV	cos φ	P_{gen} MW	P_{min} MW	P_{max} MW	Q_{min} p.u.	Q_{max} p.u.
1	GNA	2	2471	17	0.85	1750	0	2100	-1	1
2	GNB	1	2118	17	0.85	1500	0	1800	-1	1
3	GNF	2	1412	17	0.85	1120	0	1200	-1	1
4	GSB	1	1765	17	0.85	1400	0	1500	-1	1
5	GSC	2	1059	17	0.85	870	0	900	-1	1
6	GSE	2	1412	17	0.85	1100	0	1200	-1	1
7	GSI	2	2118	17	0.85	1500	0	1800	-1	1

Tab. A.7: Overview of synchronous machine parameters (adapted from [382])

		Unit	Machine type	
			Type 1	Type 2
Power plant name		-	GNB, GSB	GNA, GNF, GSC, GSE, GSI
Inertia constant (rated to S_n)	H	s	4.612	5.7059
Damping coefficient	D	p.u.	0 - 2	0 - 2
Synchronous reactance	x_d	p.u.	2.52	2.16
	x_q	p.u.	2.4	2.03
Transient reactance	x'_d	p.u.	0.287	0.22
	x'_q	p.u.	0.287	0.22
Subtransient reactance	x''_d	p.u.	0.1909	0.17
	x''_q	p.u.	0.1909	0.18
Transient time constants	T'_d	s	0.55	1.08
	T'_q	s	0.5	0.5
Subtransient time constants	T''_d	s	0.025	0.018
	T''_q	s	0.025	0.018

Tab. A.8: Speed governor parameters

K p.u.	T_1 s	T_2 s	T_3 s	T_4 s	P_n p.u.	P_{min} p.u.	P_{max} p.u.
5	0.5	0.1	0.95	0.8	$P_{n,turb}=P_{n,g}$	0	1

Tab. A.9: IEEE type AC4 excitation system parameters

T_r s	T_b s	T_c s	K_a p.u.	T_a s	K_c p.u.	$V_{i,min}$ p.u.	$V_{i,max}$ p.u.	$V_{r,min}$ p.u.	$V_{r,max}$ p.u.
0.02	12	1	200	0.04	0.2	-0.2	0.2	-4	4

Tab. A.10: DC-branch data

Name	Bus i	Bus j	l km	U_n kV	R‘ p.u.	L‘ p.u./s	C‘ p.u.·s
NE-SC	NC	SC	1000	640	0,005	0	0

Tab. A.11: VSC-HVDC converter base data

Name	S_n MVA	U_{c_DC} kV	U_{n_DC} kV	U_{n_AC} kV	Droop MW/p.u.	P_{min} MW	P_{max} MW	Q_{min} p.u.	Q_{max} p.u.
VSC NC	900	620	320	400	0.005	-900	900	-900	720
VSC SC	900	620	320	400	0.005	-900	900	-900	720

Tab. A.12: VSC-HVDC converter extended type data

Name	Mode DC[5]	Mode AC[6]	P_{VSC} MW	Q_{VSC} MVAr	V_{VSC} p.u.	r_{VSC} p.u.	x_{VSC} p.u.	b_{Shunt} p.u.	r_{Tr} p.u.	x_{Tr} p.u.
VSC NC	1	1	-400	110	1	1e-4	0.164 3	0	5e-4	0.112
VSC SC	2	1	400	0	1	1e-4	0.164 3	0	5e-4	0.112

Tab. A.13: Parameters of VSC-HVDC converter inner control loop

Name	Description	Value [362]	Estimated value [420]
$K_{p,Id}$ / $K_{p,Iq}$	Proportional gain d-/q-axis PI controller	0.48	55.28
$K_{i,Id}$ / $K_{i,Iq}$	Integrator gain d-/q-axis PI controller	149	0.32

[5] Converter control mode DC: 1 = constant power, 2 = DC slack, 3 = DC droop
[6] Converter control mode AC: 1 = PQ, 2 = PV

Tab. A.14: Parameters of VSC-HVDC converter outer control loop

Name	Description	Value [362]	Chosen value
$K_{p,P}$	Proportional gain of active power controller	0	0.01
$K_{i,P}$	Integrator gain of active power controller	33	33
$K_{p,Q}$	Proportional gain of reactive power controller	0	0.1
$K_{i,Q}$	Integrator gain of reactive power controller	33	33
$K_{p,Udc}$	Proportional gain of DC voltage controller	8	0.1
$K_{i,Udc}$	Integrator gain of DC voltage controller	272	272
$K_{p,Uac}$	Proportional gain of AC voltage controller	0	0
$K_{i,Uac}$	Integrator gain of AC voltage controller	30	30
G_{Droop}	Dynamic Droop Constant	-	11.8

A.6 Previously published papers and articles

Parts of the scientific results of your work, which are presented coherently in this work, have already been previously published in international publications. For the sake of transparency and scientific probity, these are listed below.

A.6.1 Peer-reviewed Journal Papers

C. Brosinsky and P. Sauerteig, "A moving horizon approach for dynamic state and model parameter estimation for high-fidelity power system simulation" under review for publication.

C. Brosinsky, R. Krebs, D. Westermann, "Embedded Digital Twins in future energy management systems: paving the way for automated grid control," in Automatisierungstechnik, vol. 68, no. 9, pp. 1–15, 2020.

A. Kummerow, C. Monsalve, **C. Brosinsky**, S. Nicolai, and D. Westermann, "A Novel Framework for Synchrophasor Based Online Recognition and Efficient Post-Mortem Analysis of Disturbances in Power Systems," Applied Sciences, vol. 10, no. 15, 2020, doi: 10.3390/app10155209.

A. M. Prostejovsky, **C. Brosinsky**, K. Heussen, D. Westermann, J. Kreusel, and M. Marinelli, "The future role of human operators in highly automated electric power systems," Electric Power Systems Research, vol. 175, 2019, doi: 10.1016/j.epsr.2019.105883.

A.-K. Marten, V. Akhmatov, T. Sørensen, R. Stornowski, D. Westermann, **C. Brosinsky**: "Kriegers Flak - Combined Grid Solution. Coordinated Cross-Border Control of a Meshed HVAC/HVDC Offshore Wind Power Grid". In: IET Renewable Power Generation 12 (13), S. 1493–1499. DOI: 10.1049/iet-rpg.2017.0792.

A.6.2 Book Chapter

C. Brosinsky, M. Karaçelebi, and J. Cremer "*Machine Learning and Digital Twins: Monitoring and control for dynamic security*", in "*Monitoring and Control of Electrical Power Systems using Machine Learning Techniques*", Editors: Emilio Barocio Espejo, Felix Rafael Segundo Sevilla, and Petr Korba, Elsevier, 2023.

A.6.3 Peer-reviewed Conference Papers

C. Brosinsky, A. Kummerow, M. Richter, A. Naumann, P. Wiest, S. Nicolai, and D. Westermann: "*The Role of Digital Twins in Power System Automation and Control. Necessity, Requirements, and Benefits*", Internationaler ETG-Kongress, Wuppertal, 2021.

A. Kummerow, D. Rösch, S. Nicolai, **C. Brosinsky**, D. Westermann, and A. Naumann: "*Attacking dynamic power system control centers - a cyber-physical threat analysis.*" In:

2021 IEEE Power & Energy Society Innovative Smart Grid Technologies Conference (ISGT). Washington, DC, USA: IEEE, S. 1-5.

A. Kummerow, S. Nicolai, **C. Brosinsky**, D. Westermann, A. Naumann, M. Richter: "*Digital-Twin based Services for advanced Monitoring and Control of future power systems*", IEEE Power & Energy Society General Meeting (PESGM), Montreal, QC, Canada, 2020

A. Kummerow, D. Rösch, C. Monsalve, S. Nicolai, P. Bretschneider, **C. Brosinsky**, D. Westermann, "*Challenges and opportunities for phasor data based event detection in transmission control centers under cyber security constraints*", in IEEE PES Powertech, Milano, 2019.

C. Brosinsky, F. Sass, T. Sennewald, R. Krebs, and D. Westermann, "*HVAC/HVDC Control Center - Test and Demonstrator System*", in International ETG Congress 2017, 2017, pp. 1–6.

C. Brosinsky, X. Song, and D. Westermann, "*Digital Twin – Concept of a Continuously Adaptive Power System Mirror*", in Internationaler ETG-Kongress, Esslingen am Neckar, 2019.

C. Brosinsky, D. Westermann, and R. Krebs, "*Recent and prospective developments in power system control centers: Adapting the digital twin technology for application in power system control centers*", in 2018 IEEE International Energy Conference (ENERGYCON), Limassol, Cyprus, 2018, pp. 1–6.

A. Kummerow, S. Nicolai, **C. Brosinsky**, D. Westermann, A. Naumann, and M. Richter *"Digital-Twin based Services for advanced Monitoring and Control of future power systems",* IEEE PES General Meeting, Montreal 2020

C. Brosinsky, T. Sennewald, R. Krebs, D. Westermann, "An Operator Assistant System for Fast and Reliable Decision Support based on a Dynamic Digital Mirror," in Cigré Session Paris 2020.

F. Sass, T. Sennewald, **C. Brosinsky**, M. Mangold, C. Heyde, S. Becher, R. Krebs and D. Westermann, "Control Center Implementation of Advanced Optimization and Decision Support Applications," in International Conference on Smart Energy Systems and Technologies (SEST), Sevilla, 2018.

E. Glende, P. Trojan, I. Hauer, A. Naumann, **C. Brosinsky**, M. Wolter and D. Westermann, "Communication Infrastructure for Dynamic Grid Control Center with a Hardware-in-the-Loop Model," in IEEE PES Innovative Smart Grid Technologies Europe (ISGT), Sarajevo, 2018.

C. Brosinsky, A. Kummerow, A. Naumann, A. Krönig, S. Balischewski and D. Westermann, "A new development platform for the next generation of power system control center functionalities for hybrid AC-HVDC transmission systems," in IEEE PES General Meeting, Chicago, USA, 2017.

A. Kummerow, **C. Brosinsky**, C. Monsalve, S. Nicolai, P. Bretschneider, and D. Westermann, "PMU-based online and offline applications for modern power system control centers in hybrid AC-HVDC transmission systems," in International ETG-Congress 2019 ETG Symposium, 2019, pp. 1–6.

A.6.4 Further Publications

C. Brosinsky, T. Sennewald, R. Krebs, D. Westermann: "*An Operator Assistant System for Fast and Reliable Decision Support based on a Dynamic Digital Mirror*". In: Cigré eSession 2020/ Cigré Session Paris 2021.

C. Brosinsky, T. Sennewald, R. Krebs, and D. Westermann, "*Applicational Concept for a Dynamic Power System Mirror in the Control Room*", in Cigré Symposium Aalborg, 2019.

C. Brosinsky, F. Sass, R. Krebs and D. Westermann, "*Dynamische Netzleitwarte ermöglicht kombinierten Betrieb von Hochspannungs-Gleichstrom- und Drehstrom-Technologie*", in Energiewirtschaftliche Tagesfragen, vol. 67, no. 5, pp. 50–53, 2017.

B. List of Abbreviations

AC	Alternating current
AI	Angle index
ARMA	Autoregressive moving average
AS	Assistant system
AVR	Automatic voltage regulator
BCU	Bay control unit
C2RT	Close to real-time
CA	Contingency analysis
CE-OH	Central European operation handbook
CGMES	Common grid model exchange specification
CIM	Common information model
CKF	Cubature Kalman filter
CPF	Continuous power flow
CPS	Cyber physical system
CR	Control room
CRC	Cyclic redundancy check
DACF	Day ahead congestion forecast
DAE	Differential algebraic equation
DC	Direct current
DDM	Dynamic digital mirror
DER	Distributed energy resources
DMS	Distribution management system
DSA	Dynamic security assessment
DSE	Dynamic state estimation
DSO	Distribution system operator
DSS	Decision support system
DT	Digital Twin
ED	Euclidean distance
EEG	Renewable Energy Sources Act/ Erneuerbare-Energien-Gesetz (german)
EKF	extended Kalman filter
UKF	unscented Kalman filter

EMS	Energy management system
ENTSO-E	European Network of Transmission System Operators for Electricity
EPRI	Electric Power System Research Institute
ERP	Enterprise resource planning
ES	Expert system
FACTS	Flexible AC transmission system
FPGA	Field programmable gate array
GLDPM	Generation and Load Data Provision Methodology
GPS	Global Positioning System
GPU	Graphics processing unit
GUI	Graphical user interface
HEO	German: Höhere Entscheidungs- und Optimierungsfunktionen / Advanced functions for decision support and optimisation
HPC	High performance computation
HV	High voltage side
HVDC	High voltage direct current
ICT	Information and communication technology
IED	Intelligent electronic device
IMM	Information model management
IoT	Internet of things
IT	Information technology
KF	Kalman filter
LV	Low voltage side
MHE	Moving horizon estimation
ML	Machine learning
NC-OS	Network code operational security
NERC	North American Electric Reliability Corporation
OMI	Other metering infrastructure
OPF	Optimal power flow
OT	Operational technology
PDC	Phasor data concentrator
PDP	Phasor data processor
PF	DIgSILENT PowerFactory
PMU/ PMUs	Phasor measurement unit / Phasor measurement units

PSA	Protection security assessment
PST	Phase shifting transformer
RAS	Remedial action scheme
RMS	Root mean square
ROCOF	Rate of change of frequency
RT	Real time
RTU	Remote terminal unit
SA	Situation awareness
SCADA	Supervisory control and data acquisition
SCC	Short-circuit calculation
SCOPF	Security constrained power flow
SE	State estimation
SIPS	System integrity protection scheme
SNR	Signal to noise ratio
SPS	Special protection scheme
SSA	Steady-state security assessment
SSE	Stationary state estimation
SSOT	Single source of truth
SystemOS	System operating system
TCP/IP	Transmission Control Protocol/Internet Protocol
TSCOPF	Transient security constrained power flow
TSO	Transmission system operator
TL	Transmission line
UDP	User Datagram Protocol
UI	User interface
UML	Unified Modeling Language
UTC	Coordinated Universal Time
VSC	Voltage source converter
WACS	Wide area control system
WAMPAC	Wide area monitoring, protection and control
WAMS	Wide area management system
WAP	Wide area protection
WLS	Weighted least squares

C. List of Symbols and Units

Notation

The notation used in this work is described below using the letter „a" as an example. All variables are represented by means of italic letters (*a*). Time-discrete values are indicated by a circumflex (â). A matrix is indicated by a bold capital letter (***A***). Transposed matrices are expressed by ($\boldsymbol{A}^T$). Vectors are represented by a bold lower-case letter (***a***). Scalar variables are represented in physical quantities as an upper-case letter (*A*) and in related quantities (per unit [p.u.]) as a lower-case letter (a). Complex variables are represented in underlined lower case ($\underline{a}$). Furthermore, the notation for a vector range is expressed as [$A_{\min}$, ..., $A_{\max}$] or alternatively as $[A]_{A_{\min}}^{A_{\max}}$ within in this work.

Symbols

Symbol	Description	Unit
$\hat{\boldsymbol{y}}^{m}$, $\hat{\boldsymbol{y}}$	Discrete measurement vector, discrete output vector	p.u.
x_n, $\dot{x}_n$	Differential state variable n and its derivative	p.u.
z_n, $\boldsymbol{z}$	Algebraic state variable n, vector of state variables	p.u.
μ	Mean	-
B, *b*	Susceptance (absolute and related value)	S, p.u.
C, *c*	Capacitance (absolute value, and related value)	p.u.
$\cos(\varphi_n)$	Nominal power factor	-
D	Synchronous machine speed damping	p.u.
G, *g*	Conductance (absolute and related value)	S, p.u.
H	Synchronous machine inertia constant	$\frac{\text{MWs}}{\text{MVA}}$
h	Numerical integration step size	s
$\underline{I}$, *I*, $\underline{i}$, *i*	Current (complex value, absolute value, and related value)	kV, p.u.
i, *j*, *n*	Applied as subscript to indicate a varying quantity of variables	-
j	Imaginary part of a complex number in the complex plane	$\sqrt{-1}$
K	Governor controller gain	-
k	Time instant	-
K_a	Exciter controller gain	s
K_c	Exciter current compensation factor	-
L, *l*	Inductance (absolute value, and related value)	p.u.
n	Rotor speed	p.u.
N	Horizon length	-

Symbol	Description	Unit
P	Active power	MW
p, $\boldsymbol{p}$	Parameter, vector of parameters	-
$p_{gov,ref}$	Controller reference power	p.u.
p_{max}	Maximum gate limit	p.u.
p_{min}	Minimum gate limit	p.u.
p_{ref}	Parameter reference value	-
Q	Reactive power	MVAr
$\boldsymbol{Q}$, $\boldsymbol{R}$, $\boldsymbol{S}$	Weighting matrices	-
r_{1d}	Rotor d-axis resistance	p.u.
r_{1q}	Rotor q-axis resistance	p.u.
r_{2q}	Rotor q-axis resistance	p.u.
S_b	Base apparent power	MVA
S_n	Rated apparent power	MVA
t	time	s
T_1	Governor time constant	s
T_2	Governor derivative time constant	s
T_3	Servo time constant	s
T_4	Water/boiler time constant	s
T_A	Synchronous machine acceleration time constant	s
T_a	Exciter controller time constant	s
T_b	Exciter filter delay	s
T_c	Exciter filter derivative time constant	s
T'_{d0}	Synchronous machine d-axis open circuit transient time constant	s
T'_{d0}	Synchronous machine q-axis open circuit transient time constant	s
T''_{q0}	Synchronous machine d-axis open circuit sub-transient time constant	s
T''_{q0}	Synchronous machine q-axis open circuit sub-transient time constant	s
T_r	Exciter measurement delay	s
$\underline{U}$, U, $\underline{u}$, u	Voltage (complex value, absolute value, and related value)	kV, p.u.
u_k	Short-circuit impedance of transformers	%
u_{kr}	Short-circuit impedance (real part) of transformers	%
u^m	Measurable input variable	-
$V_{i,max}$	Exciter controller maximum input	-
$V_{i,min}$	Exciter controller minimum input	-
$\boldsymbol{x}$, $\dot{\boldsymbol{x}}$	State vector, and derivative of a state vector	p.u.

Symbol	Description	Unit
X, x	Reactance (absolute value, and related value)	Ohm
x_{1d}	Rotor d-axis parameter	p.u.
x_{1q}	Rotor q-axis parameter	p.u.
x_{2q}	Rotor q-axis parameter	p.u.
x_{ad}	Mutual reactance d-axis	p.u.
x_{aq}	Mutual reactance q-axis	p.u.
x_{d}	Synchronous machine d-axis synchronous reactance	p.u.
x'_{d}	Synchronous machine d-axis transient reactance	p.u.
x''_{d}	Synchronous machine d-axis sub-transient reactance	p.u.
x_{fd}	Rotor d-axis field damper winding reactance	p.u.
x_{q}	Synchronous machine q-axis synchronous reactance	p.u.
x'_{q}	Synchronous machine q-axis transient reactance	p.u.
x''_{q}	Synchronous machine q-axis sub-transient reactance	p.u.
$\underline{Y}$	Admittance matrix	p.u.
Y	Admittance	Ohm
y^{DT}	DT model output variable	-
y^{m}	Measurable output variable	-
Z, z	Impedance (absolute value, and related value)	Ohm
δ	Rotor angle	rad
θ	Voltage angle	rad
σ	Variance	-
φ	Phase angle	radians
ψ	Flux linkage	p.u.
ω_{n}	Nominal angular frequency	s^{-1}
w	random Gaussian noise	p.u.
χk	Vector of differential states and guessed parameters	-

Units

Unit	Description
%	Percent (fraction of hundred)
A	Ampere
d	Day
dB	Decibel
F	Farad
h	Hour
k	Kilo (metric prefix 10^3)
M	Mega (metric prefix 10^6)
Mbps	Megabyte per second
p.u.	per unit (relative value, expressed as fraction of a defined base unit)
s	Second
S	Siemens
V	Voltage
VA	Volt Ampere
VAr	Volt Ampere reactive
W	Watts

D. List of Tables

E. List of Figures